Vacuum Science and Equipment. 26th Conference with International Participants

IOP Conference Series: Materials Science and Engineering
Volume 781

Sudak, Russia
15 - 22 September 2019

Editors:

S.B. Nesterov
L.L. Kolesnik

ISBN: 978-1-7138-2666-8
ISSN: 1757-8981

Printed from e-media with permission by:

Curran Associates, Inc.
57 Morehouse Lane
Red Hook, NY 12571

CURRAN ASSOCIATES INC. proceedings .com

Some format issues inherent in the e-media version may also appear in this print version.

Institute of Physics
Dirac House, Temple Back
Bristol BS1 6BE UK

Phone: 44 1 17 929 7481
Fax: 44 1 17 920 0979

techtracking@iop.org

Additional copies of this publication are available from:

Curran Associates, Inc.
57 Morehouse Lane
Red Hook, NY 12571 USA
Phone: 845-758-0400
Fax: 845-758-2633
Email: curran@proceedings.com
Web: www.proceedings.com

TABLE OF CONTENTS

IOP Conf. Series: Materials Science and Engineering **781** (2020) 011001 doi:10.1088/1757-899X/781/1/011001

Main results of the 26th conference with international participants Vacuum Science and Equipment

S.B. Nesterov[1], L.L. Kolesnik[2]

Russian Vacuum Society, 105005, Moscow, Radio str., 12
Electronic and Vacuum Technologies LLC, 105005, Moscow, Radio str., 12

E-mail: info@vacuum.org.ru

Abstract. In this preface the main results of the 26th conference with international participants Vacuum Science and Equipment that was held on 16-22 September 2019 are discussed.

This volume publishes the materials of the XXVI scientific and technical conference international participants Vacuum Science and Equipment.
The main directions of the conference:

- physical phenomena in a vacuum; vacuum systems and devices.
- calculation and modeling of vacuum systems and technological processes;
- special technological equipment;
- thin-film coatings and nanostructures;
- vacuum technologies and the aerospace complex;
- biomedical technologies;
- emission processes and sources of charged particles in devices and devices;
- history of vacuum technology.

From 1994 to 2019. under the auspices of the Russian scientific and technical vacuum society. Academician S.A. Vekshinsky was organized and held 26 conferences Vacuum Science and Equipment.
They took place in the Moscow Region (1994), Gurzuf (1995-1999), Sudak (2000-2005, 2011, 2012, 2015-2019), Adler (2006, 2007, 2009, 2010), Dagomys (2008), Gaspra (2013), Feodosia (2014).
The XXVI conference is dedicated to the 150th anniversary of the opening of the periodic system of elements of D. I. Mendeleev.

We thank the following organizations for their help:

- Vakuummash JSC
- Electronic and Vacuum Technologies LLC
- Intek Analytics JSC

IOP Conf. Series: Materials Science and Engineering **781** (2020) 011002 doi:10.1088/1757-899X/781/1/011002

Peer review statement

All papers published in this volume of *IOP Conference Series: Materials Science and Engineering* have been peer reviewed through processes administered by the proceedings Editors. Reviews were conducted by expert referees to the professional and scientific standards expected of a proceedings journal published by IOP Publishing.

IOP Conf. Series: Materials Science and Engineering **781** (2020) 012002 doi:10.1088/1757-899X/781/1/012002

Development and researches of vacuum gauges with sensitive elements created by MEMS technology

R E Kuvandykov and R A Teteruk

Saint-Petersburg, "D.I. Mendeleev Institute for Metrology", Moskovskiy avenue, 19

E-mail: ruskuw@mail.ru

Abstract. The article is devoted to the development and research of sensitive elements, transducers and vacuum gauges created by MEMS technology and intended for measuring absolute pressure in a wide range from 10^{-3} to 10^{5} Pa.

Currently in the sphere of low absolute pressure (vacuum) there is a number of problems, at first instance, connected with commercial unavailability of domestic high precision measuring facilities for measuring low absolute pressure. The following major problems can be outlined.

1. Lack of domestic comparison standard for carrying out collation between state primary standard GET 49-2016 and GET 101-2011, carrying out inter-laboratory comparisons, and also carrying out international collations.
2. Lack of domestic serially produced high precision measuring facilities conforming to the standard of 1 and 2 category by GOST 8.107-81. Since 2008 up to the present moment only two domestic produced vacuum gauges have passed trials for type approval, while their metrological characteristics do not meet requirements for the standards of 1 and 2 category [1].
3. Sanction restrictions, placed by Western countries, do not allow to procure a number of high precision measuring facilities of foreign production.

To solve mentioned problems D.I. Mendeleev Institute for Metrology proposed to develop domestic vacuum gauges, meeting the following specified requirements:

- Compact size of sensitive elements;
- Low dependency on the type of gas;
- Domestically produced components;
- Low cost.

Analysis showed that those requirements are most fully met by construction of sensitive elements produced by technology of microelectromechanical systems (MEMS). The results of research work, held proactively in the department of state standards in the sphere of pressure measurement [1], in 2017 allowed to start in D.I. Mendeleev Institute for Metrology experimental design work «Research and development of high precision deformation devices for measuring low absolute pressure in the range $1 \cdot 10^{-3}$ - $1 \cdot 10^{4}$ Pa, created by technology of microelectromechanical systems".

It was decided to develop two types of prototype models: vacuum gauge, based on resonance method of pressure measurement and vacuum gauge, based on membrane-capacitance method using compensation and compression. At the stage of compiling technical design specification, requirements for metrological characteristics of the developed prototype models were formulated as follows (see Table 1).

Table 1 – Characteristics of prototype models

Name of metrological characteristics	Value of metrological characteristics
Prototype model type 1	
Range of measurements	from 10 Pa to 10^4 Pa
Relative measurement error	not more than ±(2…1) %
Prototype model type 2	
Range of measurements	from 10^{-3} Pa to 10 Pa
Relative measurement error	not more than ± (10 -2) %

Sensitive element of vacuum gauge (picture 1), with the principle of action based on measuring gas spring rigidity (resonance type), its construction represents a plate on the hangers, placed between fixed planes.

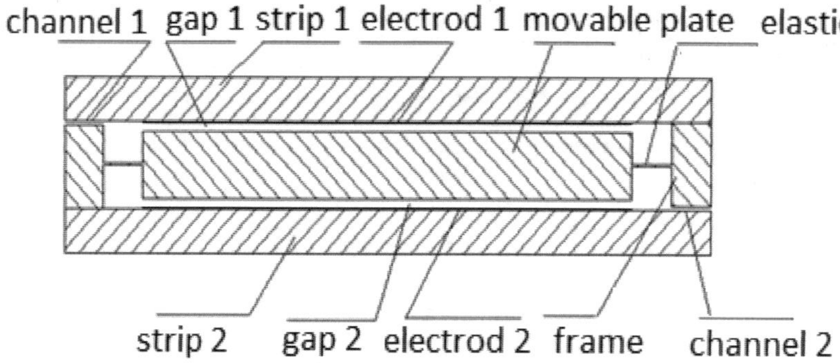

Picture 1 – draft of sensitive element of resonance type

Gas spring is formed by electrodes plane №1 and №2, and also by siliceous movable plate as shown on picture 2.

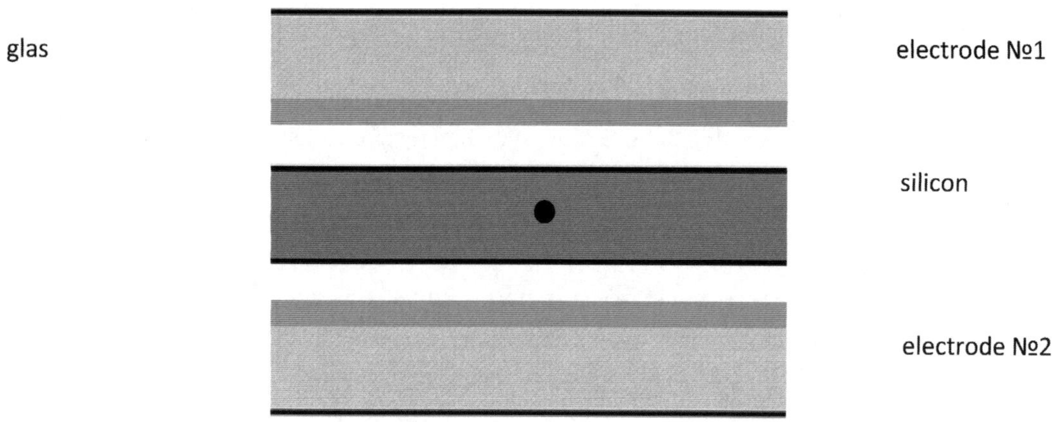

Picture 2 – Construction of sensitive element of resonance type

Metrologically significant dimensions of sensitive element of the resonant type are shown at picture 3.

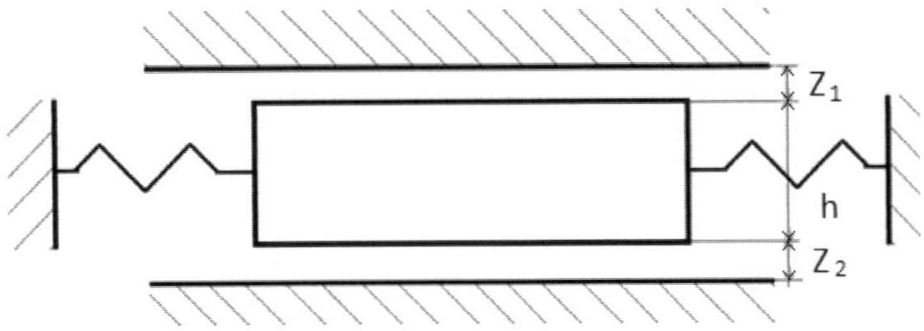

Picture 3 – Metrologically significant dimensions of sensitive element

where: h - thickness of movable plate;
Z_1 - reference value of gap (gas spring) №1;
Z_2 - reference value of gap №2.

Resonant frequency of plate vibration depends on the gas spring rigidity.

$$f_p{}^2 = \frac{G_M+G_n}{4*\square^2*m}, \tag{1}$$

where G_M- mechanical rigidity of a plate, G_n- pneumatic rigidity of a gas spring, m – weight of a movable plate.

Electrical scheme of pressure transducer of the resonance type is shown on picture 4.

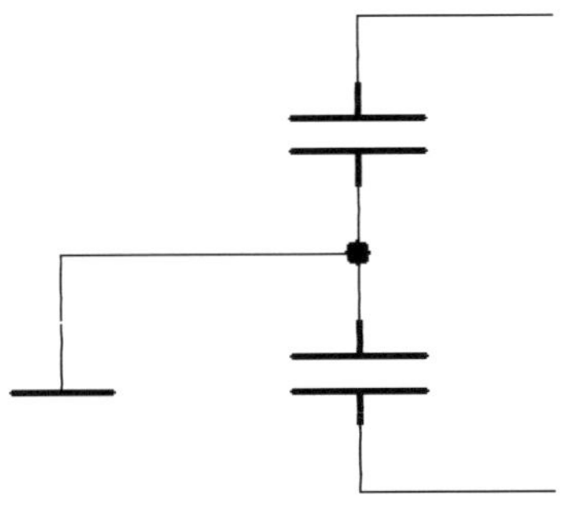

Picture 4 – Electrical scheme of pressure transducer of the resonance type

In general case parameters of plate movement are described by the following equation:

$$m*\frac{d^2y}{dt^2}+B*\frac{dy}{dt}+G*y-\frac{\square*\square_0}{2*(Z+y)^2}*\left(U_1{}^2-U_2{}^2\right)*S=0, \tag{2}$$

where y=a*sin(ω*t)- functional relation of plate movement, a- amplitude of vibration, ω- circular frequency of vibration, B- coefficient of viscous friction power of gas environment, G− $rigidity,$ ε – dielectric permeability of gas environment, ε_0 – dielectric constant, U_1 *u* U_2 – electrical voltage on electrodes.

The dependence of resonance frequency of plate movements on the gas pressure can also be obtained by the method of small displacements, if equate amount of energy, stored in gas spring, to

IOP Conf. Series: Materials Science and Engineering **781** (2020) 012002 doi:10.1088/1757-899X/781/1/012002

changes in vibrating system energy. In this case the dependency of measured gas pressure on the frequency of plate vibrations can be presented as follows:

$$P = 2 * \square^2 * \square * h * Z * f^2, \quad (3)$$

where ρ – density of plate material.

Dependence of the pressure on squared frequency of movements is linear:

$$P = K * f^2, \quad (4)$$

where K=$2 * \square^2 * \square * h * Z$ – coefficient of conversion, can be calculated from material density and value of geometrical dimensions of a movable plate.

Picture 5 shows an experimental dependence of pressure on squared frequency, received in the process of prototype model trial.

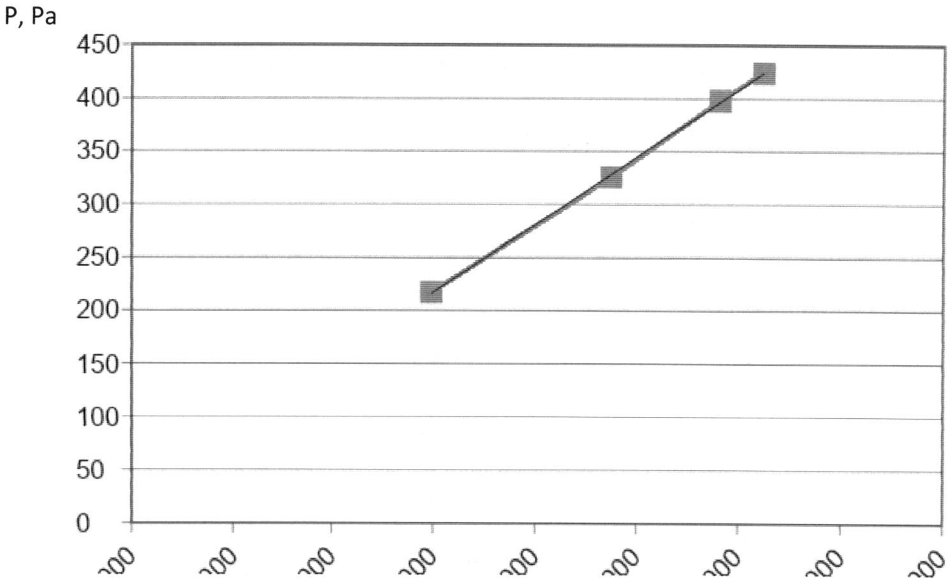

Picture 5 – Experimental dependence of pressure on squared vibration frequency

The functional relation of approximation showing that tangent of inclination angle of the straight line equals 3,18e-4 $\frac{Pa}{Gh^2}$, is also displayed on the diagram. This value coincides with earlier calculated value of the coefficient of conversion K.

IOP Conf. Series: Materials Science and Engineering **781** (2020) 012002 doi:10.1088/1757-899X/781/1/012002

Picture 6 – Sensitive element of the resonance type

Picture 7 - Pressure sensor made by technology MEMS

The second type of a sensitive element, created by MEMS technology, is based on gas pressure measuring by electrostatic pressure compensation and in its construction represents a membrane with plane-parallel distributed electrodes (picture 8).

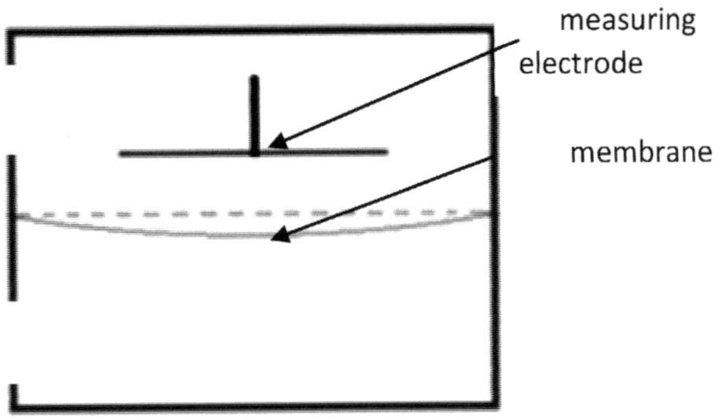

Picture 8 – Construction of a compensative transducer

Membrane deformation, created when supplying measured gas pressure, is compensated by electrostatic negative pressure (while providing constant voltage to the measuring electrode). The formula of conversion is as follows:

$$P = \frac{\varepsilon_0}{2 * Z^2} * U^2, \quad (5)$$

where ε_0- dielectric constant, h – value of gap between membrane and measuring electrode plane, U – value of constant voltage, at which compensation of membrane deformation took place.

The third type of sensitive element, created by MEMS technology, in its construction is similar to the sensitive element of the resonance type, with the difference that not gas spring rigidity is measured, but the value of power loss (work against force of friction on the gas side) with vibratory movement of a plate. The principle of action of the sensitive element of this type is illustrated on picture 9.

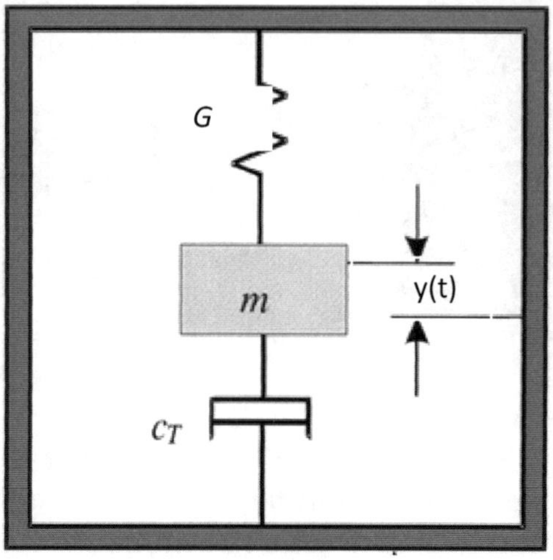

Picture 9 – Principle of measuring the value of power loss with vibratory movements of a sensitive element plate

Composite plate with weight m on mechanical hangings with rigidity κ makes forced vibrations $z(t)$, approaching by its form to harmonic. In the process of movement the plate plane experiences power on the side of gas, that is proportional to absolute gas pressure. To maintain certain amplitude of forced vibrations it is required to replenish energy spent for work against power on the side of gas. The energy is replenished by electrostatic actuator $y(t)$. The amount of replenished energy is proportional to absolute gas pressure.

The results of vacuum gauge calibration, using sensitive element, embodying the principle of measuring energy loss, are shown on picture 10.

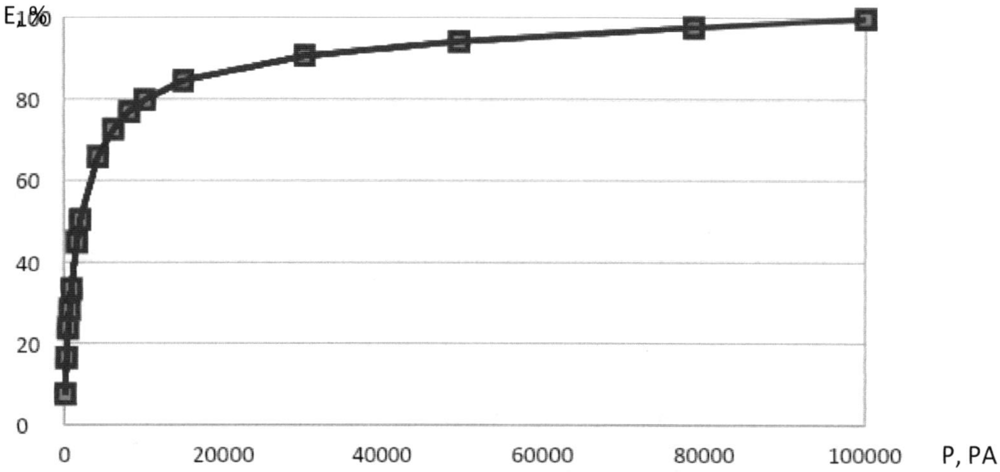

Picture 10 – Dependency of energy loss on pressure

On the x- axis the pressure in Pascal is displayed, on the y- axis – energy loss in conventional units (%) is displayed. It should be mentioned that with the increase of pressure energy loss increases non-linear.

Picture 11 shows a diagram which allows to estimate sensitivity of the method (the first derivate of calibration curve) depending on the absolute gas pressure.

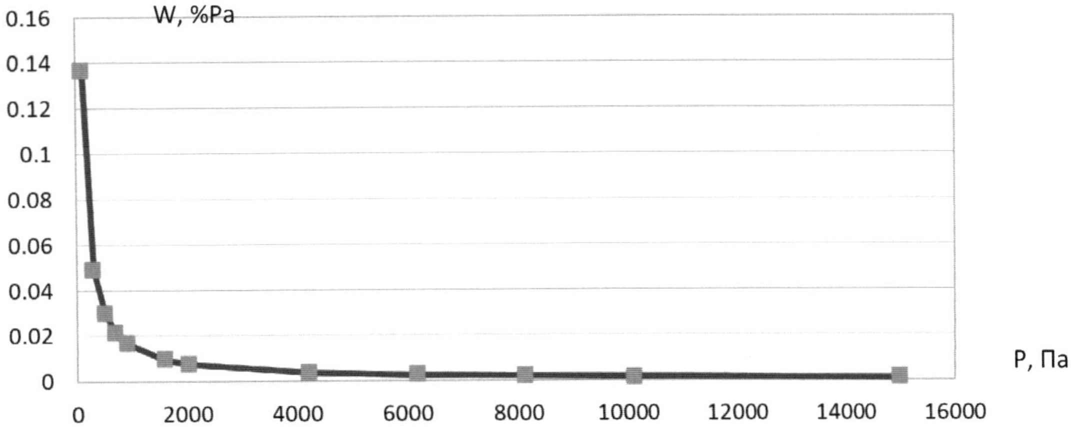

Picture 11 – Sensitivity of the method in dependence on absolute gas pressure

It is evident from the diagram that sensitivity decreases by 2 points under the pressure over 10000 Pa. The results of vacuum gauge calibration, actualizing method of measuring energy loss, were held with the use of standard vacuum gauge MKS Baratron and are represented in table 2.

Table 2 – Results of vacuum gauge calibration

Value of MEMS-vacuum gauge, Pa	**Value of standard, Pa**	**Relative error, %**	**SKO, %**
60.68	60.64	0.1	0.6
91.7	91.81	-0.1	0.4
111.6	111	0.6	0.3
301.5	301.7	-0.1	0.1
600.3	599.5	0.3	0.4
897.4	893	0.4	0.3

The external appearance of vacuum gauge with transducers, created by MEMS technology, is shown on picture 12.

Picture 12 – Vacuum gauge of the resonance type

The results of the research allow to deduce an inference, that developed models satisfy the requirements of technical design specification, and allows to continue work on creation of serially produced wide-frequency-band precision vacuum gauge, using several sensitive elements at once, created by technology MEMS and actualizing various principles of measuring absolute pressure.

References

[1] V.N. Gorobey, A.Y. Garshin, R.E. Kuvandykov. “Resonant subatmospheric transducer, created on technology MEMS” Vacuum technique and technology -2017. Contributions to 24-й All-Russian scientific-technical conference with international participation 06.06.2017.

IOP Conf. Series: Materials Science and Engineering **781** (2020) 012003 doi:10.1088/1757-899X/781/1/012003

Metrological support of test leak in vacuum

D M Fomin

Saint-Petersburg, D.I. Mendeleev Institute for Metrology, Moskovskiy avenue, 19

E-mail: vacuum@vniim.ru

Abstract. The article discusses work principles of test leaks in vacuum, metrological support and main problems arising during operation.

Test leaks are widely used in various industry fields for calibration of equipment used to control leak tightness and emission of gas in vacuum systems and produced items. For example, parameters of mass spectrometer helium leak detector are adjusted by means of test leak (helium leak) installed inside the device (or in the tested vacuum system).

The following types of test leaks (helium leaks) are currently approved as measurement tools and are included in the Federal informational fund on uniform measurements provision (State Registry of Measurements Tools). See Table 1.

Table 1 – Test leaks (helium leaks), included in State Registry of Measurements Tools

N	Name of Measurement Tools	Reproduced flow rate range, $Pa \cdot m^3/s$	Maximum permissible relative error of flow reproduction, %
1.	Gelit 1	$7 \cdot 10^{-10}$ - $2 \cdot 10^{-8}$	±15
2.	Gelit 2	$3 \cdot 10^{-11}$ - $7 \cdot 10^{-10}$	±20
3.	10xxxx	$1 \cdot 10^{-9}$ - $9 \cdot 10^{-6}$	±15
4.	Fx4xxx	$1 \cdot 10^{-10}$ - $3 \cdot 10^{-5}$	±15

Test leaks allow adjusting leak detector in order to get correct quantitative characteristics of leak tightness. Before mass spectrometer helium leak detectors were included into Federal informational fund on uniform measurements provision, qualitative characteristics of leak detectors (including maximum permissible relative error) were determined with their help.

The main equation used in mass spectrometer helium leak detectors to calculate the rate of inleakage Q, has the form [1]

$$Q = Q_{ЭТ} \cdot \frac{\alpha_{изм.} - \alpha_{ф}}{\alpha_{изм.эт.} - \alpha_{ф}} \quad (1)$$

where $Q_{ЭТ}$ – the amount of flow of helium test leak, installed inside the leak detector or in the vacuum system, specified in the certificate on verification (calibration) of leak;

$\alpha_{изм.}$ – the rate of tracer gas flow in vacuum system or in produced item, measured by leak detector in real time;

$\alpha_{изм.эт.}$ – the rate of tracer gas flow from the leak, installed inside the leak detector (or in vacuum system), measured by leak detector;

IOP Conf. Series: Materials Science and Engineering **781** (2020) 012003 doi:10.1088/1757-899X/781/1/012003

α_{ϕ} – background signal.

As can be seen from the formula, the test leak makes a significant contribution in accuracy and reliability of test results, therefore metrological support of this equipment is essential to improve the accuracy of tests results.

The most common test leaks are diffusion and capillary ones. Operation principle of diffusion test leaks is better to be considered by the example of test leak (helium leak) Gelit 1 and Gelit 2. The difference in construction modifications is in the diverse material of permeable element – quartz glass (Gelit 1) and molybdenum glass (Gelit 1).

Diffusive test leak (picture 1) is a hermetically sealed metallic tank (1) with a pipe (3) on one side for the connection with the tested vacuum system or vacuum system of leak detector. Inside the tank a tube is connected with the permeable element being a ball-shaped flask (2) made of molybdenum or quartz glass. The walls of the tank, tubes and permeable element create enclosed volume filled with helium up to the pressure from 6 to 100 kPa. Diffusion of helium through the flask walls of permeable element takes place during the operation. Wall thickness, surface area of permeable element as well as helium pressure in the tank define the range of produced flow.

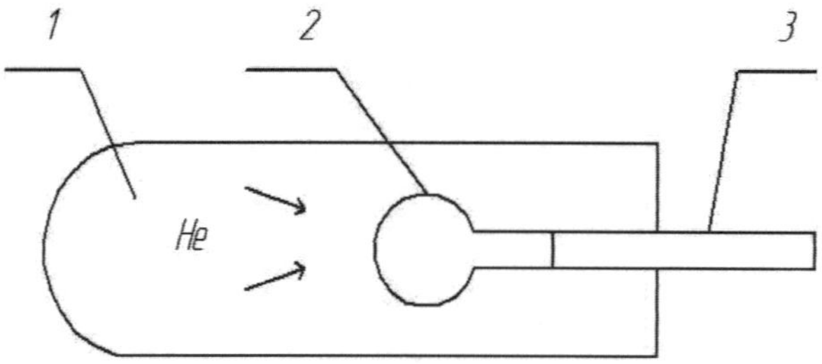

Picture 1. Diffusion test leak

Capillary test leak has similar construction but instead of permeable element, a capillary made of different materials (glass, fluoroplastic, metal, etc.) is installed. Tracer gas is filled in under the pressure of 0.1 to 1 mPa, in some cases pressure reaches 10 mPa. The major advantage of capillary test leaks is the possibility of filling them with different types of gas, which do not pollute a capillary and do not react with its surface.

In Russian Federation at present time the check gauge of the highest accuracy for test leak verification (calibration) is the State reference (operational) check gauge of the unit of gas flow in the vacuum in the range of 1,0·10-12...1,00 Pa·m3/s (GVET 49-2-2006). Total inaccuracy of gauge measurements is characterized by the sigma estimate of the result of measurements (S∑0): in the range of 1·10-12 to 1·10-9 Pa·m3/s not more than 0,1-0,015; in the range of 1·10-9 to 1 Pa·m3/s not more than 0,015. The check gauge has three measurement units with different operational principles – liquid-mechanical measurement unit with the range of measures 1·10-7 to 1 Pa·m3/s; cumulative measurement unit with the range of measures 1·10-9 to 1·10-4 Pa·m3/s; reduction gear-box measurement unit with the range of measures 1·10-12 to 1·10-5 Pa·m3/s. The check gauge is kept and used in D.I. Mendeleev Institute for Metrology.

The main problem in the usage of test leaks is the difference of the temperature of the measured flow and the temperature of test leak itself. The necessity of temperature correction application in the flow rate produced by test leaks is caused by the fact that, as a rule, the temperature, at which test leak is operated, does not coincide with the temperature of its calibration. [2]

In the research [2] the temperature correction is proposed as follows:

$$Q_T = Q_K \frac{T}{T_K} e^{-\frac{E}{R}(\frac{1}{T}-\frac{1}{T_K})} \qquad (2)$$

where Q_T – flow produced by test leak at the temperature T, $\frac{Pa \cdot m^3}{s}$;

Q_K – flow produced by test leak at the temperature of its verification (calibration) T_K, $\frac{Pa \cdot m^3}{s}$;

T, T_K – absolute temperatures, K;
E – activation energy for permeable material of test leak element, cal/ mole;
R – universal gas constant.

To specify corrections with the help of this formula it is necessary to identify the rate of activation for permeable elements of test leak (for example, quartz or molybdenum glass and fluoroplastic).

Currently manufacturers most often indicate average calculated value of temperature correction for specific test leak model. As a rule, the value of temperature correction is 3%/K. Manufacturers also indicate the flow rate in measuring units and the temperature of verification (calibration). To recalculate the flow the following formula should be used:

$$Q_T = Q_K(1 + \frac{K_T}{100})^{(T-T_K)} \quad (3)$$

where Q_T – flow produced by test leak at the temperature T, $\frac{Pa{\cdot}m^3}{s}$;

Q_K – flow produced by test leak at the temperature of its verification (calibration) T_K, $\frac{Pa{\cdot}m^3}{s}$;

K_T – temperature correction indicated by manufacturer;
T – operating temperature of test leak, K;
T_K – temperature of verification (calibration) of test leak, K.

However, the temperature can be disregarded if the flow rate is set as a molecular consumption, i.e. penetration of amount of substance (mole) per unit of time. [3] In our country it is not common yet to delineate flow in measuring units mole/s, however, world vacuum community has taken a decision for gradual transition to measuring flow (especially for standard equipment) namely in mole/s measuring units, which allows to neutralize temperature dependence while measuring.

To measure precisely with the help of test leak it is necessary to use device for temperature stabilization (aerial or aquatic thermostat, heat shield etc.). It is experimentally proven in the article [2], that for diffusion test leak type Gelit 1 and Gelit 2 15 minutes exposure under the newly set temperature is enough for temperature stabilization of the measured (reproduced) flow.

The next important problem during the operation of the test leak is the impact of background flows. Background flows have significant impact on the accuracy and reliability of measuring results. These flows are caused by two main reasons – background flows of measuring system (of the leak detector, vacuum system itself and so on) and background flows, which come from connection pipe walls of the test leak. The first type of flows can be successfully dealt with, for example, through system heating, cleanup, prolonged unloading, and also “zeroing” of the background signal with the help of modern devices of leak testing.

The second type of background signal arises due to improper storage of test leak – long-time storage in the air with the opened connection pipe, the impact is even more in case of long time storage of the test leak with connection pipe closed by vacuum valve. In this case tracer gas is accumulated in permeable element and the flow rate for some indefinite time can significantly rise from 30% to 100% in comparison with assigned rate. Unfortunately, modern leak detectors as a rule produce rather low vacuum at the input, that does not allow promptly to pump out tracer gas accumulated in permeable element, which creates additional error when measuring by means of this test leak.

It is experimentally proven that the longer is the unloading of test leak connection pipe, the higher is the accuracy of measured flows and the lower is the measuring error.

Proceeding from facts mentioned above it is very important to comply with correct terms of test leak storage, and especially before verification (calibration), as improper storage can significantly raise high the flow rate for indefinite time, as a result the operator can get incorrect nominal rate of the flow.

Another serious problem is elaboration and serial domestic production of test leaks with an option to be filled with various gases (hydrogen, nitrogen, argon, Freon, etc.). These test leaks are required for calibration of sniffer leak detectors. It is also necessary to conduct state testing for the approval of the type for determination of metrological characteristics of test leaks data. For the current moment

IOP Conf. Series: Materials Science and Engineering **781** (2020) 012003 doi:10.1088/1757-899X/781/1/012003

there are no test leaks with the option of filling by another than He4 gas approved as SI in RF State Register.

In the year 2019 OJSC "Vaktron" on demand of D.I. Mendeleev Institute for Metrology elaborated and produced new capillary test leaks – helium (nitrogen) leak VAKTRON VKT with the option of filling both by helium and nitrogen.

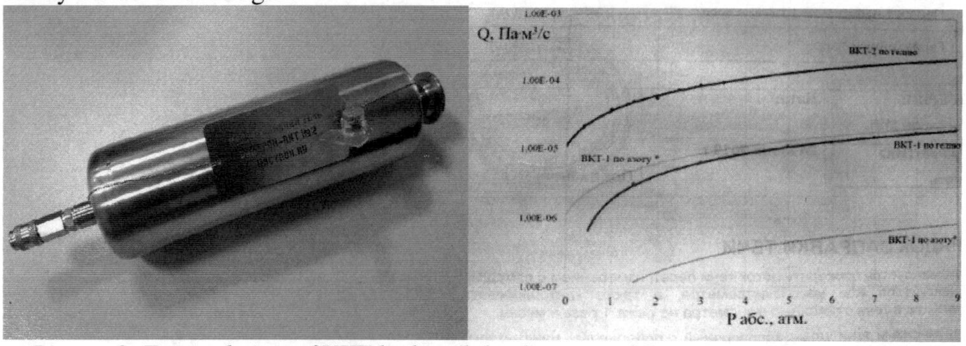

Picture 2. External view of VKT leak and the diagram of flow dependence on the pressure inside the leak tank

The main advantage of new test leaks is an opportunity to regulate the flow by increasing or reducing test gas pressure inside the tank. Discharge of test leak can occur as to vacuum, so to the atmosphere. It is also worth mentioning about the simplicity of filling the tank with tracer gas individually. Preliminary tests confirmed characteristics indicated by manufacturer.

Timely metrological service and also compliance with operating and storage rules for test leaks in vacuum will allow to significantly enhance accuracy of measurements with the help of this equipment, and therefore improve the quality of produced items, reduce the probability to make wrong decision concerning tested objects.

References

[1]. Kuzmin V.V., Alyaev V.A. Vacuum measuring technique: monograph – Kazan: Publishing house: Kazan state technological university, 2009

[2]. Afanasieva L.A., A.I. Evlampiev, I.V. Tvorogov Metrological certification of diffusive helium calibrated leaks // Vacuum technique and technology. Volume II, № 1, 1993

[3]. Handbook of Vacuum Technology. Second edition. Edited by Karl Jousten, 2016

IOP Conf. Series: Materials Science and Engineering **781** (2020) 012004 doi:10.1088/1757-899X/781/1/012004

Prepare of nanocomposite structures through the magnetron deposition of material on the colloidal silica films and investigation of their properties

E N Galaganova and D A Sotnikov

Bauman MSTU, str. Baumanskaya 2, 5/1, Moscow

lenagalaganowa@mail.ru, denis.sotnikov.2012@mail.ru

Abstract. The paper presents a study of nanocomposite structures prepared by magnetron sputtering of titanium on a template formed during the deposition of colloidal silica particles on a metal substrate. The effect of power and spraying time on the formation of nanocomposite structures was studied. The features of the processes of growth of a metal film on an opal matrix and the Implantation of metal into its structure are revealed.

1.Introduction

The use of nanocomposite materials in micro- and nanoelectronics devices allows to discover new results, properties and regularities of physical and chemical phenomena. The area of application of nanocomposites is quite wide: nanocomposite InN (quantum dots) – semiconductor nanoelectronics devices; C, Si, Ge(3D- nano-optical systems) – active systems for amplification, generation, laser control ; Ga, Ti, Fe (3D – nanocomposites and nanograting, metallic photonic crystals – magnetic memory elements, reflectors.[1] In technology not only direct photonic crystal structures are applied, but also inverse photonic ones formed using the base of microsphere colloidal particles, acting as a template.

The inverse photonic crystal formation is impossible without the implantation process of functional material into template. The characteristics and properties of inverse structure change depending from the functional material used. For example, using magnetic materials as the implantation ones you can prepare porous 3-D structure with the fill factor up to 76%.

The «Luntic» (Figure 1) thin film deposition machine is for the research conduction in the field of thin film deposition by the magnetron sputtering and vacuum-evaporation technique. The thermal vacuum method of thin film's production is based on the heating of substance in vacuum until it's active evaporation and condensation of vaporized atoms on the surface of the substrate. The advantages of the method are the high purity of the deposited material (the process is carried out at high and ultrahigh vacuum), versatility (films of metals, alloys, semiconductors, dielectrics are applied) and relative ease of implementation [2]. One of the main disadvantages of thermal method is a droplet phase. The droplet phase is formed due to the appearance of vapors in the evaporated material, that captures macroscopic particle passing through a melted metal [3]. Also, the method is characterized by a non-adjustable deposition rate and deposition energy, variable and uncontrollable deposition energy [4].

The magnetron method consists in cathode sputtering with high-energy ions of working gas. The magnetron sputtering method is characterized by localized plasma near the cathode that allows to increase the sputtering rate significantly through the high ion current density at lower operating pressure [4].

In the process of forming nanocomposite materials, one of the main factors is the energy of the particles of the functional material. The energy determines the effectiveness of its insertion into template [5]. The film growth rate and adhesion are significantly increased when using a magnetron sputtering system, since the kinetic energy of the spraying atoms significantly exceeds the kinetic energy of the evaporated ones [3]. Based on the above factors, the method of magnetron sputtering is chosen as the method of spraying functional coatings.

Figure1. Magnetron Sputtering equipment

An experiment of introducing titanium into template formed by sedimentation from a colloidal solution of silica was carried out with the installation of magnetron vacuum deposition [6-7].

2. Investigation of the influence of technological deposition modes of functional material on the degree of template filling

Possible variants for the titanium deposition on the template in the upper layer are presented in Figure 2.

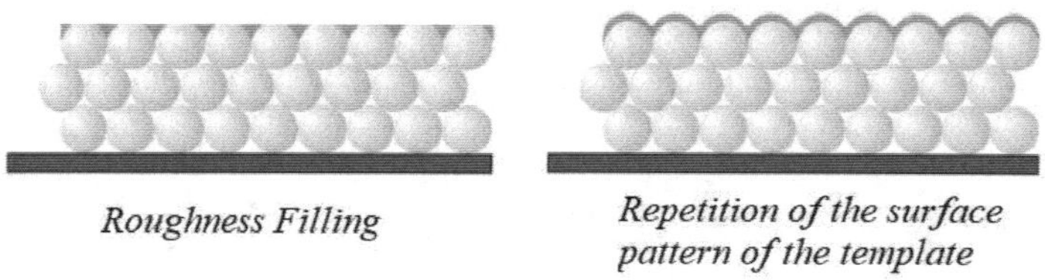

Figure 2. Filling irregularities Top layer formation variants

The following experiment was carried out to discover features of the growth process of metallic film on the opal matrix: on the template formed on the Al2O3 aluminum substrate deposited titanium films by different thickness at different energies of the sprayed particles by the magnetron sputtering. (Fig.3) Table 1 shows the main parameters values of the titanium deposition mode. Thus, titanium with a thickness of 50 and 100 nm was deposited on a smooth test sample on a template, at each power value (700 W and 350 W).

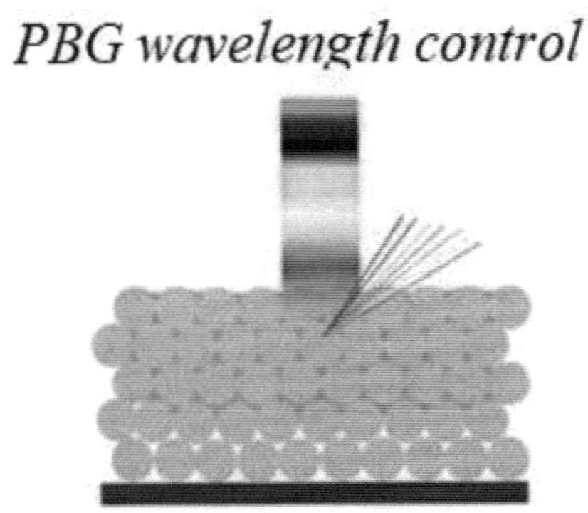

Figure 3.- Template depth control

Table 1 – Technological mods

Parameter name	Symbol	Unit of measurement	Value
Vacuum chamber pressure	P	Pa	2
Argon consumption	Y	cm^3/min	55
Magnetron power	Pw	W	700 350
Current strength	I	A	1,43 1,2
Voltage	U	B	401 292
Duration	t	sec	60/30
The distance to the substrate	L	mm	65

For these four samples, spectrophotometric analysis was carried out before and after titanium deposition (Fig. 4 - 5).

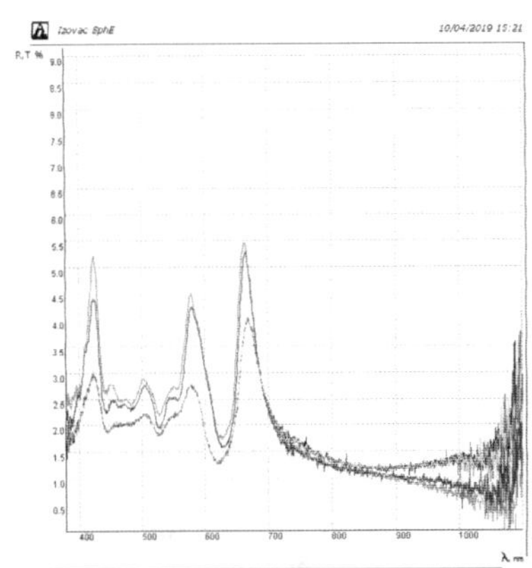

Figure 4 Spectra before titanium deposition

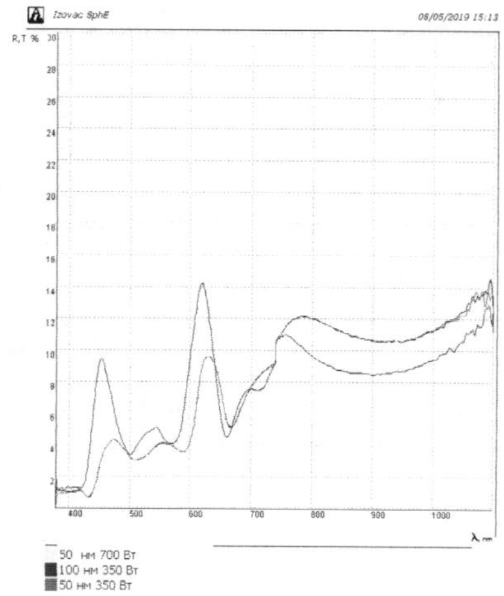

Figure 5. Spectra after titanium deposition

The number of layers in the template structure is calculated by the formula (1) and is equal to 5. The photonic band gap of the sample without a titanium film was observed at wave-lengths of 430 nm, 500 nm, 570 nm, and 650 nm. Position dates of the photonic band gap for the samples after the depositing titanium are in the datasheet 2.

$$h = \frac{\lambda_1}{2 \cdot n_{eff} \cdot \cos\theta (1 - \frac{\lambda_1}{\lambda_2})}, \qquad (1)$$

n_{eff} - effective refractive index of the material.

Table 2 –PBG position dates

Thickness 50 nm power 350 W			
Peak 1	Peak 2	Peak 3	Peak 4
471 nm	588 nm	632 nm	786 nm
Thickness 50 nm, power 700 W			
436 nm	521 nm	663 nm	1093 nm
Thickness 100 nm, power 350 W			
450 nm	540 nm	617 nm	750 nm
Thickness 100 nm, power 700 Вт			
There is no photonic band gap			

There are characteristic peaks on the reflection spectra of samples with a titanium thickness of 50 nm that is equal to 350 W power. This indicates the photonic crystal structure of the samples is unchangeable. The forth sample shows a film of 100 nm thickness formed at the 700w power that shaped titanium blanket coating on a template and there is not found PBG.

The shift in the position of all the PBG peaks in the first three samples toward infrared waves also indicates the filling of the inter-spherical space of the template, since when titanium is introduced into the silica film, a composite structure is obtained with an eigenvalue of the refractive index, which can be determined by the formula. It proves the result reached in work [8], which is devoted to the assessment of the uncertainty of measuring the fill factor of the opal matrix by the embedding material.

$$n_{eff} = \sqrt{n_{op}^2 \; f_{op} + n_{air}^2 \; f_{air} + n_{imp}^2 \; f_{imp}} \quad (2)$$

n_{op} – n_{op} - opal refractive index, f_{op} - f_{op} – opal matrix fill factor , n_{air} n_{air} – - air refractive index, f_{air} – f_{air} - matrix fill factor, n_{imp} – n_{imp} - refractive index of the implant material, f_{imp} – f_{air} - matrix fill factor.

On the contrary, when spraying at a power of 700 W, titanium begins to actively deposit on the top layer and forms a blanket film on the template, due to which the material does not reach the layers located below.

3.Conclusion

The results of theoretical and experimental studies have shown that for the introduction of functional material, it is necessary to use the method of magnetron sputtering at low power (300-350 W), as it provides penetration of the material to a depth of 3-4 layers of photonic crystal structure.

References

[1] Bulygina E. V., Makarchuk V.V., Panfilov YU. V. Nanorazmernye struktury: klassifikaciya, formirovanie i issledovanie: Uchebnoe posobie dlya Vuzov. – M.: SAJNS-PRESS, 2006.– 80s.

[2] Gol'dade V.A., Fedosenko N. N. Metody polucheniya tonkih plenok. GGU im. F. Skoriny, 2015. – 41s.

[3] Dunyushkina L.A. Vvedenie v metody polucheniya plenochnyh elektrolitov dlya tverdooksidnyh toplivnyh elementov: monografiya / L.A. Dunyushkina. – Ekaterinburg: URO RAN, 2015. – 126 s.

[4] Panfilov YU. V. Nanesenie tonkih plenok v vakuume/ Tekhnologii v elektronnoj promyshlennosti, №3, 2007.

[5] F.I.Grigor'ev Osazhdenie tonkih plenok iz nizkotemperaturnoj plazmy i ionnyh puchkov v tekhnologii mikroelektroniki: Uchebnoe posobie / Mosk. gos. in-t elektroniki i matematiki. M., 2006.

[6] Kuleshova V. L., Panfilova E. V., Prohorov E. P. Automated device for vertical deposition of colloidal opal films //2018 International Russian Automation Conference (RusAutoCon). – IEEE, 2018. – C. 1-5.

[7] Sapoletova, N. A. ., Martynova, N. A. ., Napol'skij, K. S. ., Eliseev, A. A. ., Lukashin, A. V. ., Kolesnik, I. V. ., ... Tret'yakov, YU. D. (2011). Samosborka kolloidnyh chastic v prisutstvii elektricheskogo polya. FIZIKA TVERDOGO TELA, 53(6), 1064-1068.

[8] Galaganova E.N. Ocenka neopredelennosti izmereniya koefficienta zapolneniya opalovoj matricy materialom vnedreniya // Sbornik VIII Mezhdunarodnoj molodezhnoj nauchnoj konferencii «Nanomaterialy i nanotekhnologii: problemy i perspektivy». - Saratov: Saratovskij gosudarstvennyj tekhnicheskij universitet imeni Gagarina YU.A., 2018. - S. 22-26.

IOP Conf. Series: Materials Science and Engineering **781** (2020) 012005 doi:10.1088/1757-899X/781/1/012005

Research of the influence of vacuum deposition on the growth rate of MoS_2 films

A I Belikov[1], V N Kalinin[1] and A V Sikov[1]

[1]Department of Electronic Technologies in Mechanical Engineering, Bauman Moscow State Technical University "BMSTU", 105005, Moscow, Russia

Email: belikov@bmstu.ru

Abstract. The research results of the working pressure and discharge power influence on the growth rate of molybdenum disulfide thin films on silicon substrates during magnetron sputtering of the target in vacuum are presented. Based on experimental data, an empirical model has been developed that relates the film growth rate to argon pressure and discharge power on a magnetron sputtering system.

1. Introduction

The magnetron deposition of thin films method is one of the widely used methods in production. This is explained by its capabilities: versatility with respect to the materials used in the process (conductors, semiconductors and dielectrics), coating intricate surfaces of products, depositing high-quality multicomponent coatings at relatively high deposition rates [1].

Thin-film coatings based on molybdenum disulfide (MoS_2) are widely used as antifriction coatings in mechanical engineering, and in recent years they have been actively studied in the field of nanoelectronics, where they can be used as a two-dimensional semiconductor. The film-forming particles deposition rate on the substrate affected onto the film structure formation peculiarities and ultimately at the thin films properties. The molybdenum disulfide thin films properties substantially depends on the thickness [2]. Ultrathin MoS_2 films (from monolayer and up to several nanometers thick) can be used in nanoelectronics (transistors, various sensors, etc.) [3]. In this regard, it is necessary to ensure the coating growth at a certain rate to obtain the desired thickness and film structure. In order to refine the technological process parameters for a new thin-film coatings formation and to efficiently choosing the technological regimes of magnetron deposition, it is necessary to take into account the influence of the main process parameters (Figure 1), such as the pressure of the working gas in the vacuum chamber, the distance between the magnetron and the substrate, and the target material, its size, and etc., on the MoS_2 films growth rate. The working gas type and its flow rate, the temperature of the substrate heating and the bias potential supplied to substrate, the substrate surface ion pretreatment presence and its parameters can also determines thin films quality. In addition, modern power supplies for sputtering systems provide stabilized sputtering modes of targets with various parameters. The gas discharge during sputtering can be maintained in constant current (DC) mode and in pulsed mode at medium frequencies (MF), with providing by the power source a frequency rate and pulse duty rate. In DC mode, the key parameters are the current and discharge power, the supplied voltage and the type of stabilized parameter (stabilization by current/voltage/power). And in the pulse mode, these parameters are supplemented with the characteristics of the pulses: frequency rate repetition, duty cycle and power in the pulse. The energy and the flow of film-forming particles reaching the substrate surface are significantly depending from the working gas pressure and the sputtering modes of the target, and the discharge power acts as the most frequently used parameter of the sputtering process. Therefore, in this work, the problem of studying and evaluating the argon pressure degree and magnetron sputtering system discharge power influence on the molybdenum disulfide thin films growth rate on silicon substrates was solved.

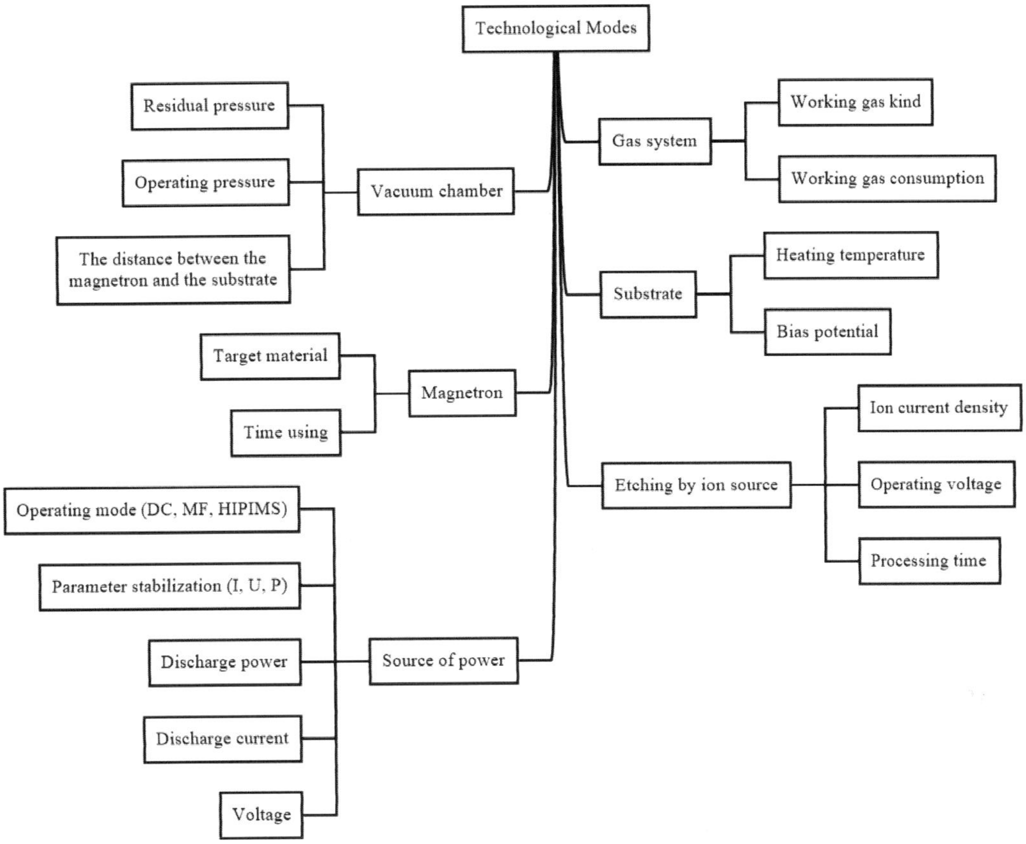

Figure 1. Scheme of technological parameters

2. Research methods

In this work, MoS_2 thin films were deposited on polished silicon substrates with a surface roughness of less than 3 nm. The thickness of the deposited films was measured by the height difference «film-substrate». For these purposes, a TR220 model profilometer was used.

After preliminary preparation of the substrate surfaces, immediately before their placement in the technological equipment chamber, masks were fixed on the substrate surfaces, which were removed after films deposition. It ensured a "film-substrate" step profile formation. By using the surface profile obtained with the profilometer, the thickness value was measured by the height difference on the prepared step. In addition, for samples with the films that were previously examined on a tribometer test, as a result of which after complete wear of the coating layer wear tracks were formed on their surfaces, from which it is also, was possible to evaluate the film thickness. The scheme of the sample with the wear tracks and the prepared line of the «step», includes plan of the points, starting from which the profilometry and the measurements were carried out, is presented in Figure 2.

At the initial measurements stage, the probe of the profilometer passed through the center line (from points 1-3) of the silicon substrate with the deposited molybdenum disulfide film (Figure 2).

The first three scans directions were carried out along the center line of the samples with the initial location of the probe at points 1, 2, and 3 and its movement to the substrate center, which made it possible with the more accuracy, constructs and estimates the complete surface profile, taking into account the profiles of wear tracks. The subsequent construction of profiles for two additional measurement areas with the scanning start from points 4 and 5 at the direction parallelly to the center line, that mades it possible to receive information of the film thickness along the «step» line. To increase the reliability 10 measurements were performed for each sample.

The thickness estimate results was carried out using both types of the results of the measurings as at the «step» lines areas and so on the obtained wear tracks profiles.

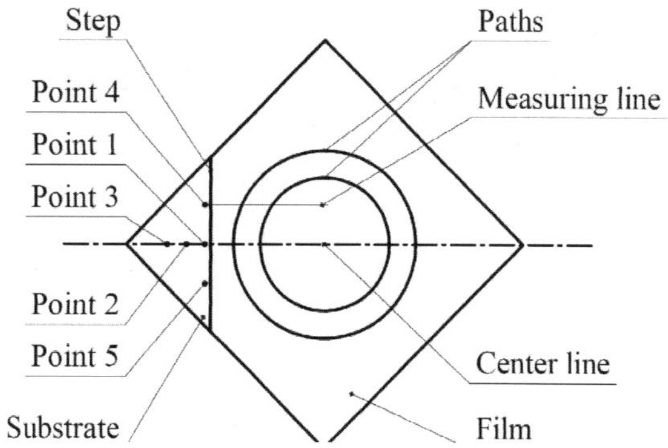

Figure 2. Scheme of the sample with the initial points for profilometry displacement

3. Results and discussion

To improve films to substrates adhesion, immediately before the molybdenum disulfide films deposition, the silicon substrates surfaces was etched throw an autonomous ion source. Ion treatment was carried out at an ion current density on the substrate of approximately 1.5 A/m^2; the treatment time was 30 minutes.

Prior to the technological processes of the molybdenum disulfide films deposition on silicon substrates, the etching depth of silicon after ion treatment was initially determined. Since the «step» was formed as a result of the overlapping of one area of the substrate with a mask (Figure 3), and the film was deposited on the area that was previously exposed with ion treatment, film growth was starting out from the surface which was formed as a result of layer etching (Figure 4).

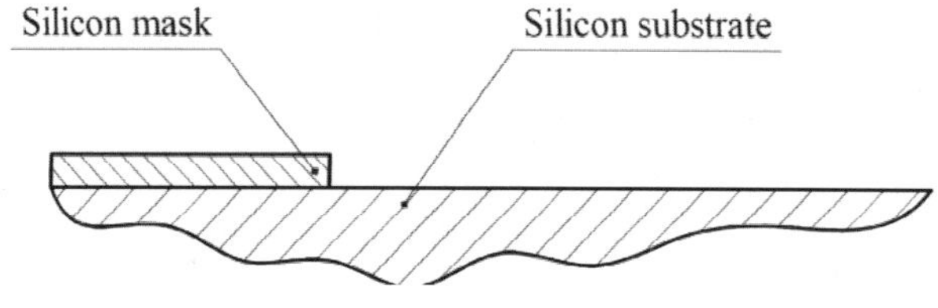

Figure 3. Scheme of the mask layout on the substrate

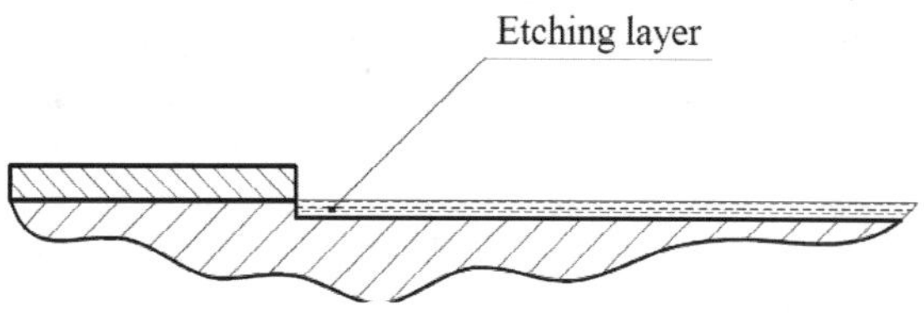

Figure 4. The substrate surface line after ion treatment

The presence of the etching effect must be taken into account when the deposited on the substrate film thickness determines since the measured value of the step profile corresponds to the measured value of h_1 (Figure 5), which does not an actual film thickness (h_2). To obtain the real film thickness h_2, the value of the etching depth must be added to the obtained value of h_1. For ion treatment that was used in this work, the measured with the profilometer etching depth value was 0.3 μm. Taking into account the obtained etching depth value, the measured values h_1 were corrected up to the film thickness values h_2.

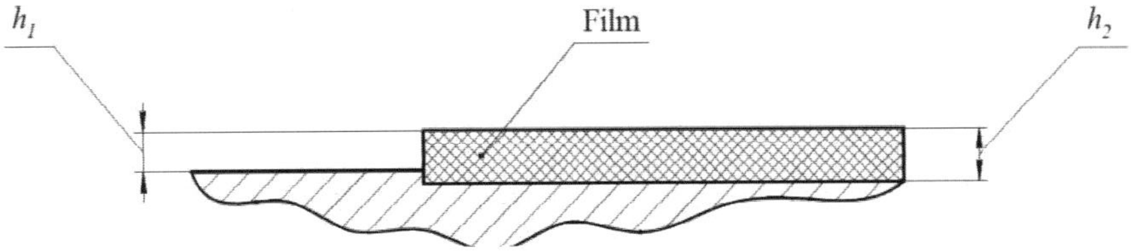

Figure 5. The film on the substrate location after removing the mask

Before loading into a vacuum chamber, the samples were pretreated in organic solvents. To degrease the surface and remove large contaminants, the silicon wafers were cleaned in an ultrasonic bath for 2 minutes at a frequency of 50 kHz. The substrates were firstly cleaned in acetone and then in isopropyl alcohol. The temperature of the solutions was 20 °C. Further, the substrates were dried, after which they were placed on a substrate holder and loaded into a vacuum chamber.

Samples with molybdenum disulfide films were prepared on a vacuum unit equipped with a magnetron sputtering system with planar targets. In this work, produced by pressing powder of DMI-7 MoS_2 powder targets were used. The diameter of the target was 75 mm, and the thickness was 5 mm.

The limiting residual pressure in the vacuum chamber of the equipment was 10^{-3} Pa. The working gas was argon. The variable process parameters were the discharge power P, (factor 1) and the argon pressure p, (factor 2).

The output parameter was the average film growth rate V, [nm/min].

The controlled and stabilized factors were: the operation mode of the power source – constant current mode, discharge current stabilization, substrate heating temperature – 250 °C, the bias potential on the substrate – the earth potential, the film deposition time – 60 minutes, the distance between the magnetron and the substrate – 100 mm.

The measurements results of the average films thickness for three series of measurements performed at different times are presented in Table 1.

Table 1 – MoS_2 thin films average thickness experimental values

P, W		p = 1 Pa			p = 0.4 Pa	
20	0.17 μm	0.14 μm	0.11 μm	0.62 μm	0.67 μm	0.74 μm
40	0.27 μm	0.24 μm	0.25 μm	0.92 μm	0.95 μm	1.00 μm
60	0.23 μm	0.27 μm	0.28 μm	1.71 μm	1.78 μm	1.83 μm
80	0.47 μm	0.45 μm	0.50 μm	2.56 μm	2.62 μm	2.67 μm

The pressure in the vacuum chamber was monitored using the AP6100 XM sensors (Pirani thermal sensor, error 15%) and AIM-X Edwards (inverse magnetron sensor, error 30%).

The operation modes of the magnetron sputtering system (power, current and discharge voltage) were set using the APEL-M-3 power supply, the values were controlled by a measurement system built into the unit with a microcontroller (error 3%).

The distance between the magnetron and the substrate was controlled by a vernier caliper (error 0.05 mm).

The deposition process time was controlled by a stopwatch (0.1 s error).

The working gas flow was installed and controlled using gas flow controllers (error 0.5%) with computer control.

The substrate temperature was maintained by a temperature controller based on feedback, the temperature sensor was a thermocouple (error 0.5 °C).

The film thickness was measured with a TR220 profilometer (error ≤10%).

For the films deposition process time (60 minutes), the average values of the film growth rate were determined (Table 2).

Table 2 – MoS_2 thin films average growth rate values

P, W	p = 1 Pa	p = 0.4 Pa
20	7.3 nm/min	16.3 nm/min
40	9.2 nm/min	20.9 nm/min
60	9.3 nm/min	34.6 nm/min
80	12.9 nm/min	48.6 nm/min

As a result of regression analysis, based on experimental data, the following mathematical model was obtained (1):

$$y = 21{,}3 + 9{,}5X_1 + 11{,}2X_2 + 6{,}7X_1X_2 \tag{1}$$

As follows from model (1), the coefficient b_2 of factor X_2 turned out to be 15.2% higher than the coefficient b_1 of factor X_1, therefore, factor X_2 (pressure) has a greater effect on the film growth rate than factor X_1 (discharge power). The combined effect of these two factors is less than the influence of only one factor X_1 by 41.8%, and less than factor X_2 by 67.2%.

Obtained results indicate the presence of a nonlinear dependence of the MoS_2 films growth rate on the power supplied to the magnetron during sputtering. Nonlinearity is observed for each of the two experimental series, in which different working gas pressure values were used. Such deviations from linearity may be explained by the formation and deepening of the sputtered region (erosion zone) on the target surface, with an increase in the time of its use, which leads to a changes in the electric fields configuration at the target surface and a changes in the trajectories of the bombarding ions relative to the target original nominal plane. Due to the presence of anisotropy in the structure of the MoS_2 crystals, a change in the angle of ions relative incidence to the axis of the crystal can lead to a change in the sputtering coefficient. Therefore, under the same sputtering conditions, by the presence of different erosion zones on the target surface, different streams of atomized material can forms.

4. Conclusion

As a result of the studies, the influence of the magnetron discharge power and argon pressure in the chamber on the growth rate of molybdenum disulfide films was established. According to the obtained mathematical model, with decreasing pressure, the film growth rate increases. An increase in discharge power also leads to an increase in the film growth rate. The influence of the pressure factor on the film growth rate was more significant.

The obtained empirical model can be used to determine the deposition modes of thin MoS_2 films with a given growth rate and obtain coatings of the required thickness by calculating the required discharge power for a specific argon operating pressure.

During the work, the etched depth of the silicon substrate after ion treatment was established, the value was 0.3 μm. This value is significant when applying thin films of molybdenum disulfide with a thickness of less than a micrometer; therefore, to increase the accuracy of measurements by profilometry methods, it is necessary to correct results by this value.

References

[1] Kiryukhantsev-Korneev F V, Horwat D, Pierson J F, Levashov E A 2014 *Letters in Journal of Technical Physics* **14** 1-2

[2] Xiaorong Gan, Huimin Zhao, Xie Quan 2016 *Biosensors & Bioelectronics* **89** 56-70

[3] Rui Cheng, Shan Jiang, Yu Chen, Yuan Liu 2014 *Nature Communications* **5** 5143

Simulation of the electrical conductivity of island-type thin films and nanostructures

O G Andreasyan[1] and S V Sidorova[1]

[1] Bauman Moscow Technical University, 2-ya Baumanskaya Street, 5/1, Moscow, Russia, 105005

Abstract. Areas of application of islet thin films in science and technology are described. Experiments on the formation of metal island nanostructures are performed. The principle of calculating the size of islands at different stages of growth is presented. The function of the tunneling current between islet nanostructures versus the deposition time is empirically deduced.

1. Introduction

The growth process of a continuous thin film can be divided into the following steps (figure 1):

1. Nucleus formation
2. Nucleus growth, islet formation
3. Coalescence of islets
4. Channel formation
5. Hole formation
6. Continuous film formation

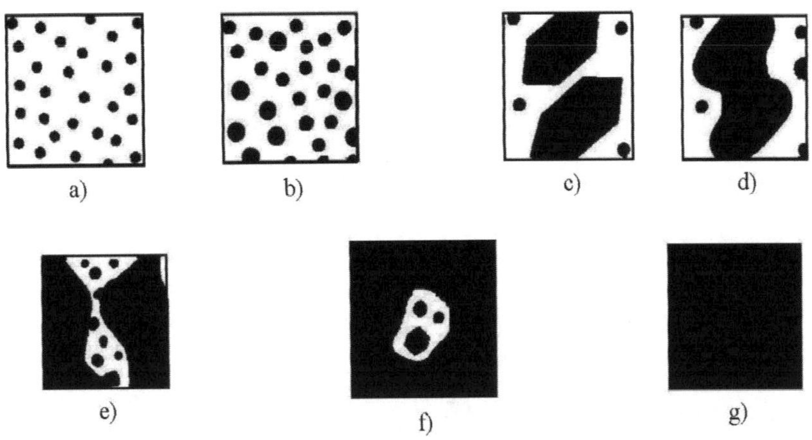

Figure 1. Stages of film growth: a – nucleus formation; b - nucleus growth, islet formation; c, d - coalescence; d - channel formation; e - hole formation; g - continuous film formation [5]

Islet nanostructures or island-type thin films (ITF) are films whose formation was completed at the stage of island formation [1].

They arouse interest in themselves because of the special properties associated with the nano-dimension of these structures in all three directions. A striking representative of such effects is the dimensional quantization of the energy levels of an electron located inside an island.

The literature discusses methods for modeling ITF. The authors show the importance of the various characteristics of these methods, because each of them has its own particular qualities, both the application and the results obtained.

Among these methods, molecular beam epitaxy (MBE), gas phase epitaxy (GPE), ion beam etching (IBE), lithography and many others are often distinguished.

One of the most promising and modern technological methods is the MBE method. Its principle of operation is based on the growth of single crystal layers due to reactions between thermally generated molecular or atomic beams of the corresponding components and the surface of a substrate located in ultrahigh vacuum at elevated temperature [1,2,3].

Scientists note that new methods for obtaining ITF can make a huge breakthrough in instrumentation at almost all of its sections. For example, ITF can be used to increase the operability of a transistor - a device that converts electrical signals [2,5-11]. If quantum dots are formed in the channel of such a device, then quantum effects such as tunneling and dimensional quantization of the energy spectrum of charge carriers can significantly improve the characteristics of the element.

Also, islet structures can be used in medicine. Some scientists believe that by applying gold nanoparticles to a surface immobilized by a dye, the sensitivity of giant Raman spectroscopy (GRS) can be significantly increased [2].

It is also indisputable that the development of methods for obtaining ITF will coordinately affect nano-electronics itself. One of the arguments is the fact that island-type structures are used as catalytic seeds to produce carbon nanotubes (CNTs) [6].

It should also be noted that quantum dots (islands of semiconductor materials 2–10 nm in size) are one of the methods for implementing qubits in quantum computers [19].

Based on the results of the literature review, it becomes clear that island-based nanostructures are an integral part of the future of mankind, and new ways of creating ITF, as well as modernization of existing ones, can play a key role in the development of technologies!

The work is devoted to the development of a mathematical model for the formation of an island-type film, with the help of which it will be possible to determine the size of structures obtained from various materials.

The aim of the study is to build a mathematical model of the conductivity of electric current through an island-type thin structure based on previously obtained experimental data.

2. Methods and materials

The method of thermal evaporation is based on the heating of substances in special evaporators to a temperature at which a noticeable evaporation process begins, and the subsequent condensation of the vapor in the form of thin films on the treated surfaces located at some distance from the evaporator [5].

When thin films are deposited in vacuum by thermal evaporation (figure 2), energy is supplied to the substance being evaporated by resistive (direct and indirect) and / or high-frequency heating, electron bombardment, heating using laser radiation or using an electron beam.

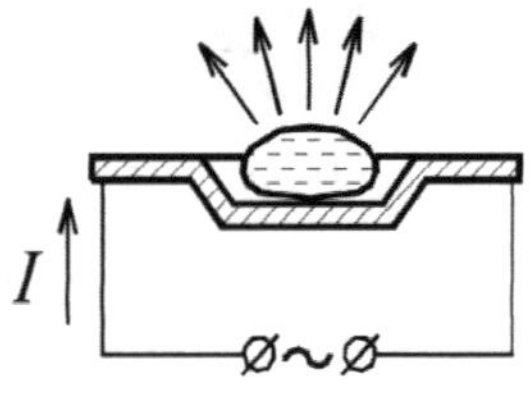

Figure 2. The scheme of the method of thermal evaporation [5]

When the temperature of the evaporated substance rises to a temperature equal to or higher than the evaporation temperature, particles of the evaporated substance leave the evaporator, are transferred to the substrate in vacuum, and condense on its surface, forming a thin film [5].

The current flows over the substrate surface either by direct tunneling (figure 3), or by tunneling through stable energy impurity states.

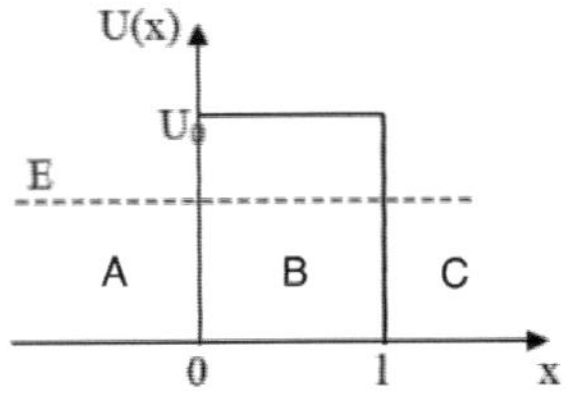

Figure 3. Particle tunneling through a potential barrier [5]

Hill also proposed a model [7], in which the participation of the substrate in the conduction mechanism between the islands was taken into account. In this case, the magnitude of the current increases, since the height of the potential barrier is less than the thermodynamic work function of the electron. In addition, the transparency of the potential barrier increases even more if there are free alkali metal ions on the surface of the substrate, which form surface electronic states in the substrate itself [8, 12].

Based on the experimental data, an empirical dependence of the current flowing on the substrate between the contacts on the film deposition time was obtained [12]. In this paper, this result will be the main one - most arguments will be based on it.

3. Vacuum processing equipment

In this work, for the formation of island nanostructures, a small-sized modular type vacuum installation was used (UVN-1M), which is intended for research in the field of thin film deposition in vacuum (figure 4).

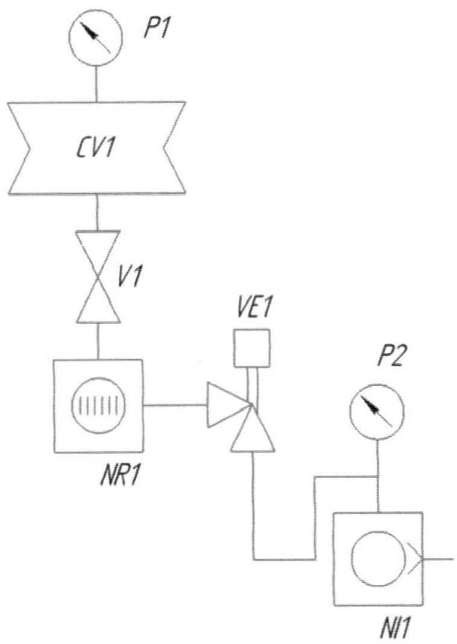

Figure 4. The basic vacuum circuit of the initial installation; P1, P2 - a wide-range sensor and a Pirani sensor, respectively; CV1 - a vacuum chamber; VE1, V1 - angular solenoid valve and butterfly valve, respectively; NR1 - turbomolecular pump; NI1 - spiral pump

Installation specifications:

- ultimate pressure: 10^{-3}-10^{-6} Pa;
- working gas pressure: 0.1–100 Pa;
- film deposition rate: 10^{-3}-15 nm/s;

• etching rate of layers: 10^{-2}-50 nm/s;
• power consumption: 1.5–5 kW;
• overall dimensions: 500×1100×1600 mm;
• weight: 90 kg.

4. Solution

The tunneling current between islands of the same material depends on the applied potential and the distance between islands [9, 10, 11]. Known view for the dependence of the density of the tunneling current:

$$j_t = e^2 V (2m\varphi)^{1/2} (4\pi^2 \hbar^2 \Delta z)^{-1} e^{-2\hbar \Delta z (2m\varphi)^{1/2}}, where \quad (1)$$

m – electron mass;
Π - electron work function of the film material;
Ћ – reduced Planck constant;
V – applied external potential;
$\otimes z$ – distance between the islands;
e – electron charge;

Let’s consider the basic experiment: a substrate was prepared from ceramic with pre-formed contact pads on it (figure 5). Let's place it in the working chamber and pump out this volume up to 10^{-3} Pa (working pressure). After heating to 80°, we begin to evaporate copper from a tungsten spiral.

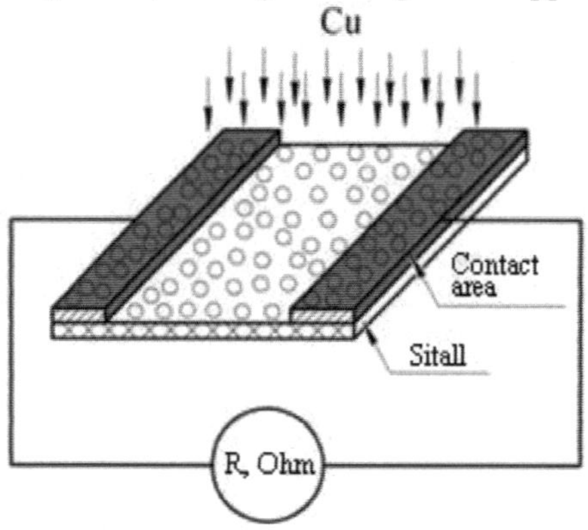

Figure 5. Substrate Connection Diagram

Then, in equation (1) we get:
φ=const;

Thus, from equation (1) we obtain:

$$j_t = AV(\Delta z)^{-1} e^{B\Delta z}, where \quad (2)$$

A= $e^2 (2m\varphi)^{1/2} (4\pi^2 \hbar^2)^{-1}$ =const>0;
B= $-2\hbar (2m\varphi)^{1/2}$ =const<0;

Let's assume that the voltage is constant (V=const), then equation (2) will simplify to:

$$j_t = \frac{C}{\Delta z} e^{B\Delta z}, where \quad (3)$$

$C = e^2 V (2m\varphi)^{1/2} (4\pi^2 \hbar^2)^{-1}$ =const;

With the growth of the film, the diameter of the islands changes according to the following law determined experimentally [6]:
$D = \{175 + 27x_1 - 22x_2 - 8x_3 + 10x_1x_2$ для $t \in [20; 35]$ $498 + 134x_1 - 59x_2 - 8x_3 + 36x_1x_2$ для $t \in [40; 65]$ $880 + 41x_1 - 14x_2$ для $t \in [70; 85]$, where

x_1, x_2, x_3 – are dimensionless magnitudes of factors: application time — x_1, evaporator – substrate distance — x_2, substrate temperature — x_3.

Interestingly, the selected time intervals are characterized by various physical processes occurring on the substrate, in terms of the number (N, pcs/μm^2) of islands and their size (D, nm): from 20 to 35 s

IOP Conf. Series: Materials Science and Engineering **781** (2020) 012006 doi:10.1088/1757-899X/781/1/012006

- *N* and *D* increase; from 40 to 65 s - *N* is stable, *D* grows; from 70 to 85 s - *N* decreases, *D* grows. Let's turn to dimensionless quantities of factors. In this case, the upper level of the factor is "1", and the lower is "–1". The coordinates of the center of the plan are zero.

From this expression we will consider only the average time interval $t\in[40; 65]$ s, since it is on this interval that the number of islands on the substrate is already stabilized and only the distance between them will affect the tunneling current [6].

Now, let's connect the distance between the islands with their diameter. Let's assume that the change in the interval between the islands is proportional to the change in the remaining free area on the substrate (figure 6).

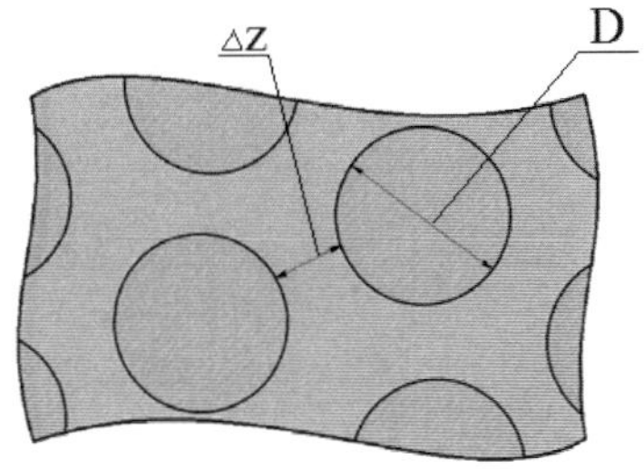

Figure 6. The process of island growth at *t*=40..65 s

Let the number of islands per unit area is N (N = const in our time interval). The area S_1 occupied by one island will be equal to:

$$S_1 = \frac{\pi D^2}{4}, where$$

D - is the average diameter of the island;

Then the area S_Σ occupied by all structures will be equal to:

$$S_\Sigma = S_1 \cdot N = N\frac{\pi D^2}{4};$$

Let the substrate area be *S*, then the time dependence of the distance between the islands is proportional to S-S_Σ. We get:

$$j_t = \frac{C_1}{S - S_\Sigma} e^{C_2(S-S_\Sigma)}, where \quad (4)$$

$C_1 = e^2 V(2m\varphi)^{-1/2}\,(F4\pi^2\hbar^2)^{-1};\ C_2 = -2F\hbar(2m\varphi)^{-1/2}$,

F - is the dimensional constant characterizing the transition from the distance between the islands to the difference between the areas

Let's us depict the dependence for the diameter for the selected interval in the following graph (figure 7).

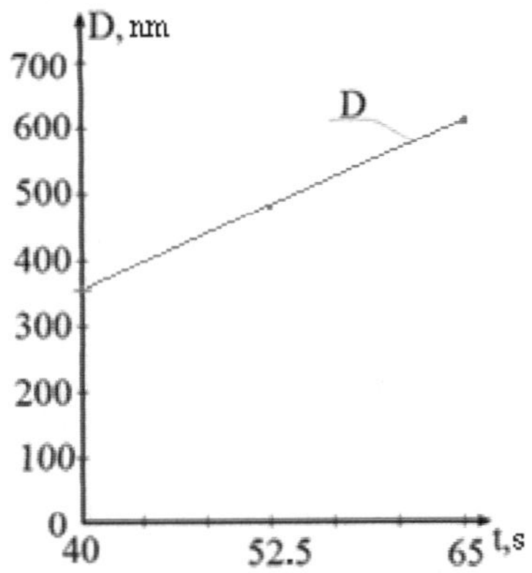

Figure 7. Graph of the island diameter versus time

Let's convert this graph to the equation:

$$D = 10.72t - 64.8\ [nm], where\ t \in [40; 65]\ s \quad (5)$$

Now, let's substitute equation (5) in equation (3) and picture the resulting graph (figure 8).:

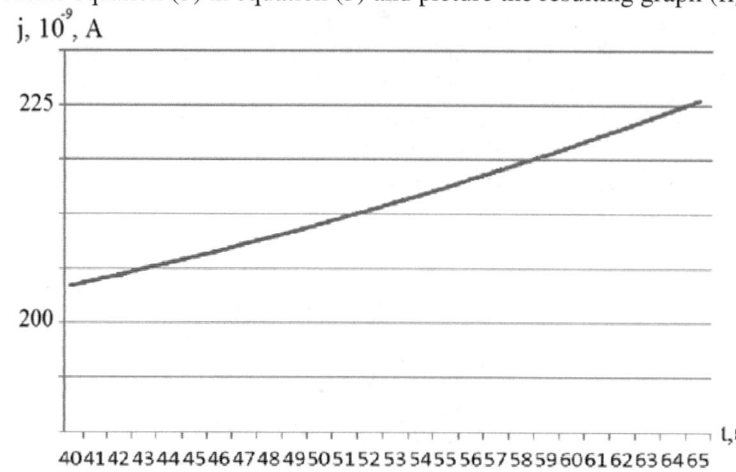

Figure 8. Characteristic dependence of current strength on time, calculated according to the model

Let's compare this dependence with that obtained in the course of a real experiment (figure 9).

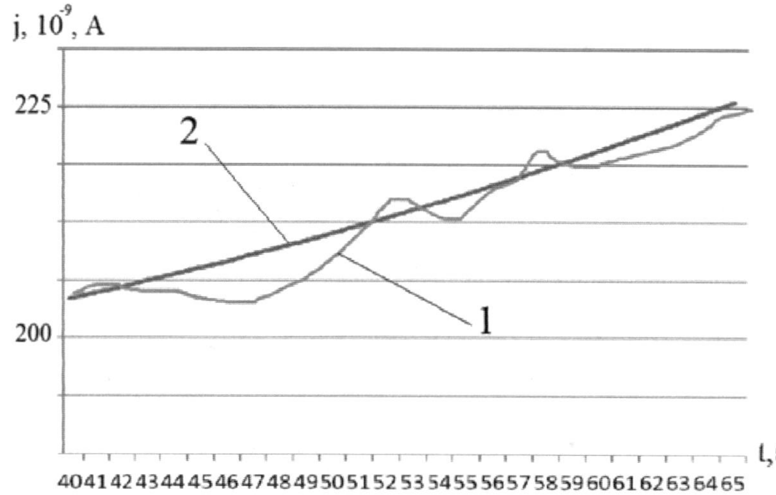

Figure 9. Characteristic dependence of current strength on time, obtained experimentally (1) and calculated according to model (2)

One of the basic assumptions of why real dependence behaves in this way may be the property of the islands themselves. The islands grow in pulses, in particular, the total surface area of two islands differs from the area of one obtained when they merge [6, 15-16]. Because of this phenomenon, "jumps" may appear inside the dependence of the tunneling current on time, since the tunneling current depends on the distance between the islands, which, in turn, is associated with their area.

The second important fact is the influence of the characteristics of the vacuum installation on the course of the experiment. In a real experiment, the compact modular vacuum installation uses the Edwards nEXT 75D turbomolecular pump as the final pump, with the help of which an operating pressure of 10^{-3} Pa is reached during the experiment. At such pressures, the concentration of molecules in the volume can be calculated from the equation of state of an ideal gas:

$$P = nkT \Rightarrow n = \frac{P}{kT}, where$$

k – Boltzmann coefficient = $1.38 \oplus 10^{-23}$ J/K;

IOP Conf. Series: Materials Science and Engineering **781** (2020) 012006 doi:10.1088/1757-899X/781/1/012006

T – working temperature = 363 K [5];

We get:

$n=2.05 \oplus 10^{17}\ pcs/m^3$

This means that during the entire experiment, quite a few gas molecules remained inside the setup, which could get on the substrate during film growth and cause sharp increases in resistances between the islands, which could cause the value of the tunneling current between them to drop.

5. Work results

As a result of the review of information sources, the advantages of instruments and devices on the islet nanostructures are considered and substantiated, the relevance of the study of ITF is determined.

A mathematical model of the dependence of the tunneling current on the geometric parameters of the ITF is developed.

6. Conclusion

The obtained mathematical model of the growth of an island thin film is idealized, since it contains some assumptions. However, even now it is capable of transmitting the correct nature of the change in the tunneling current during the formation of the ITF. In the future, the model will be finalized. Using it, it will be possible to determine not only the time of formation of islands from different materials, but also the parameters of the resulting substrate – island thin film system.

References

[1.] Sidorova S V 2010 Review of formation methods and growth modeling of islet nanostructures in vacuum *Collection of works of third All-Russian conference of young scientists and specialists "The future of Russian engineering* 142

[2.]]Borzyak P G and Kulupin Yu A 1980 *Electronic processes in islet metal films* Kiev: Naukova Dumka 240

[3.] Maysell L and Glang R 1977 *Technology of thin films* Moscow: Sovelovsk radio volume 2 chapter 2 p 768

[4.] Sidorova S V 2016 Calculation of technological modes and selection of equipment parameters for the formation of island thin films in a vacuum Moscow: BMSTU

[5.] Trusov L I and Kholmyansky V A 1973 *Islet metal films* Moscow: Metallurgy p 320

[6.] Smirnova K I 2007 Thin films in microelectronics: Textbook Tomsk: Tomsk State University of Control Systems and Radioelectronics p 94

[7.]]1978 *Physics of Thin Films* ed Hass G, Francombe M and Hoffmann R Moscow: Mir Volume 8 p 359

[8.] Chopra K L 1972 *Electric phenomena in thin films* Moscow: Mir p 435

[9.] Burshtein E and Lundqvist S 1973 *Tunneling phenomena in solids* Moscow: Mir p 424

[10.]Pestovsky Yu S 2011 Obtaining islet gold films for highly sensitive immunoassay Moscow: Moscow State University

[11.]Sidorova S V and Yurchenko P I 2011 Studies of the formation of islet nanostructures in vacuum *Nano- and microsystem technology* Moscow No 5

[12.]John G and Simmons J 1963 *Appl. Phys.* volume 34 p 238

[13.]]Palatnik L S and Ilyinsky A I 1968 Mechanical properties of metal films *UFN* volume 95 release 4 pp 613–645

[14.]Palatnik L S, Fuchs M Ya and Kosevich V M 1972 *The mechanism of formation and substructure of condensed films* Moscow: Nauka p 320

[15.]Palatnik L S and Papirov I I 1971 *Epitaxial films* Moscow: Nauka p 480

[16.]]2014 *Quantum theory in 30 seconds* ed Clegg B tr Perfiliev O Russia: Helvetica p 155

IOP Conf. Series: Materials Science and Engineering **781** (2020) 012007 doi:10.1088/1757-899X/781/1/012007

Actuator Based on a Dielectric Elastomer with Quartz as a Filler for Vacuum Technology

A M Bazinenkov[1],D A Ivanova[1],A P Rotar'[1], V P Mikhailov[1]

[1]Bauman Moscow State Technical University, 2-ya Baumanskaya, 5-1, Moscow, 105005, Russian Federation

anastasia.rotari15@gmail.com, ambazinenkov@bmstu.ru, ivanova_d_a@bk.ru, mikhailov@bmstu.ru

Abstract. Modern equipment requires precise movement of objects both at atmospheric pressure and in vacuum conditions. Moreover, vacuum equipment is subjected to a number of external disturbances that adversely affect the manufacturing process and have a direct impact on the quality of manufactured products. A promising development in this area is actuators based on dielectric elastomers, characterized by a short reaction time (milliseconds) and a simple and cheap implementation. Successful implementation depends on an understanding of its behavior and properties, depending on the choice of matrix and filler materials. Most often, expensive barium titanate is used as a filler. This paper discusses the use of quartz as a cheaper filler.

1. Introduction

Modern mechanical-engineering equipment increasingly requires precise movement of objects both at atmospheric pressure and in vacuum conditions. To ensure accurate movement and vibration isolation of objects, mechanisms of precise movements (actuators) based on "smart" materials and, in particular, controllable magneto- and electroactive materials, liquids and polymers, have become widespread [1].

Polymers that exhibit a change in shape or volume when stimulated by an electric voltage or current are called electroactive polymers, or EAPs. These active substances can be classified in accordance with their principle of operation [2, 3] as ionic and electronic ones. Ionic EAPs: the movement of ions during electrical activation is responsible for a change in shape or volume. Ions controlled by an external electric field, as a rule, diffuse through a liquid or solid electrolyte and, thus, cause an electromechanical change in the material configuration. Electronic EAPs: electric charges are displaced upon activation of electronic EAPs. The resulting electrostatic forces lead to electromechanical changes in the shape or volume of the material [4].

Dielectric elastomers (DEs) are one of the most promising representatives of electronic electroactive polymers, since they can provide large deformations and withstand high loads. The properties of this material, enclosed between the electrodes, are controlled by an external electric field: an electric voltage of several kilovolts is applied to two conductive electrodes, and the dielectric layer is compressed as differently polarized electrodes are attracted to each other. In addition, electrostriction occurring in a dielectric medium contributes to the general deformation of DE [5].

2. Main components of DE

The main components of DE are: elastomeric matrix and a dielectric powder filler. When choosing the material of the DE matrix, it is important to consider the following requirements: high dielectric constant, a wide range of operating temperatures, high electric strength, low Young's modulus; therefore, synthetic or natural rubbers are used as the DE matrix, and the matrix can also be made by mixing one, two or more materials. The mixing of several materials as components of the matrix is made in order to improve its characteristics, but often the enhancement of one parameter leads to a deterioration of other characteristics of the material [6, 7].

The DEs with matrices based on three types of polymers: silicone, acrylic and polyurethane, are considered most promising today, but other materials, for example, natural rubber, are practically not investigated due to lower values of required characteristics [8].

Table 1. Electrical and mechanical characteristics of matrix materials

Matrix material	Young's modulus, MPa	Dielectric constant	Electric strength, kV/mm	Temperature range
Silicone	0.1…10	2…5	50…350	-100…200°C
Acrylic	0.6	4…6	45…250	-10…80°C
Polypropylene	17	6…7	160	-

The DE can be improved by adding fillers to the elastomer matrix; their dielectric properties are high and can enhance the dielectric properties of elastomers several times. Therefore, finely dispersed barium titanate powder, which may have a relative dielectric constant of up to 3000, is most often used as a filler [9].

The manufacturing of a highly dielectric and finely dispersed powder of barium titanate is a time-consuming and expensive operation, so the goal of the work was to test a sample in which a cheaper and more available dielectric, quartz, was used as a filler.

Mixing the matrix material and the filler material in the correct proportions is of no small importance for the properties. When mixing, a strong correlation between the dielectric constant and electric strength should be taken into account: an increase in dielectric constant is usually accompanied by a decrease in electric strength [10].

The choice of material for the electrodes is the main problem associated with the manufacturing of actuators based on DE. This is due to the fact that they must not only withstand high electric fields (up to hundreds of kV/mm), but they must also be well compatible and maintain good electrical contact during deformation. The most obvious version of the electrodes – two metal plates, is possible, but their huge disadvantage is the high rigidity of such electrodes, which usually limits the maximum achievable deformation value [11, 12, 13].

3. Manufacturing of quartz-based DE

Based on an analysis of [5, 6, 7, 8, 9, 11], in which studies of the influence of DE components' materials on its properties are presented, silicone compound was used as a matrix for the manufacturing of prototypes, since it has the best set of properties: high dielectric strength (27 kV/mm), high elastic modulus (10 MPa) and a wide range of operating temperatures (from -60 to 200°C), as well as elongation at break up to 165% [13].

As a filler, quartz powder with a fineness of 0.5 μm was used. To study the effect of the filler concentration in the matrix volume on the DE properties, samples were prepared with a ratio of the matrix material to the filler material of 1:2, 1:4.

The manufacturing process was performed at the MT11 subdepartment of BMSTU. At the first stage, the surface of the quartz particles was covered with an active layer of polymethylsiloxane to prevent the appearance of particle agglomerates in the polymer. The obtained powder is added to the silicone compound pre-mixed with the hardener. After fifteen minutes of mixing, the suspension was placed for 20 minutes in a desiccator, which provides a pressure of 10^2 Pa using a rotary vane pump, in order to degas the polymer volume and prevent the porosity of the final product. After degassing, the polymer composition was cast into a mold providing a height of 1 mm, previously coated with a surfactant layer, and placed into a drying cabinet at a temperature of 150°C for 60 minutes. The polymerization occurred due to the chemical reaction of the vinyl group compound in the composition of silicone rubber and the hydride group in the composition of the polymerizer. After cooling the obtained polymer, samples of 40 mm in diameter were knocked out using a metal pointed form [14, 15, 16].

4. Design of quartz-based DE

The main criterion for choosing the design of DE is the direction of action. The DE is characterized by two directions of deformation: longitudinal – parallel to the field (change in thickness) and transverse – orthogonal to the field (change in surface area). The operating principle is the material volume constancy: uniaxial pressure load is equal to the biaxial tension load in the remaining spatial directions [16].

IOP Conf. Series: Materials Science and Engineering **781** (2020) 012007 doi:10.1088/1757-899X/781/1/012007

The classic version of DE for use in the actuator is a "capacitor" design. The "capacitor" design (Fig. 1) is identical to the design of a plane capacitor, but with a DE clamped between a pair of electrodes. A wide range of matrix, filler and electrode materials is suitable for this design [17].

Figure 1. Actuator design according to the "capacitor" scheme

The multilayer structure (Fig. 2), in turn, consists of actuators of the "capacitor" design connected in series. The number of layers installed affects the linear size of the actuator and, therefore, the absolute stroke when moving [17, 18].

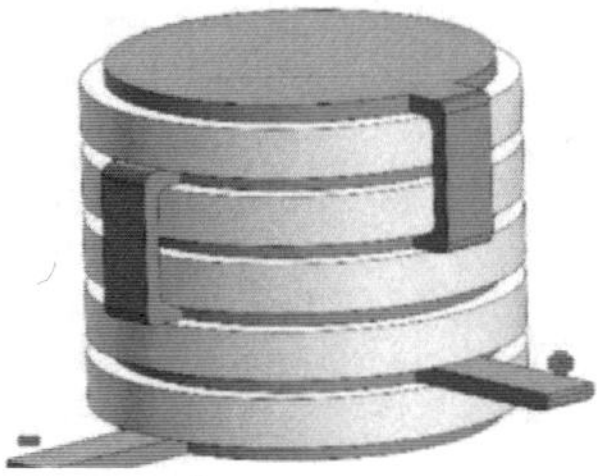

Figure 2. Multilayer actuator design

Any DE materials can also be used, but it is recommended that the electrodes be made in the form of flexible thin metal ones, since electrodes based on carbon lubrication can lead to the displacement of the layers relative to each other. Therefore, electrodes were made from rolled copper sheets using laser cutting in the form of two ribbons with round segments [19, 20].

The assembly of a multilayer DE with this electrode design was performed by sequentially bending electrode segments in perpendicular planes [21].

8 DE samples were made: one-, two-, four- and eight-layer with 60% and 80% quartz concentrations.

5. DE-based precise movement mechanism

Precise movement mechanisms are characterized by such indicators as the movement error and range, load capacity, the number of degrees of freedom and the response time, and requirements for these indicators increase over time [22].

When developing the experimental design of the actuator based on DE, it is necessary to take into account the following factors: the design should be universal and suitable for using both a single-layer and multi-layer elastomer with a different number of layers, it should be made of dielectric material to provide insulation for the purpose of protection against high voltages, it should have high rigidity, backlashes should be excluded and ease of manufacture should be provided [25].

The design of precise movement mechanism presented in Fig. 3 was developed based on these requirements.

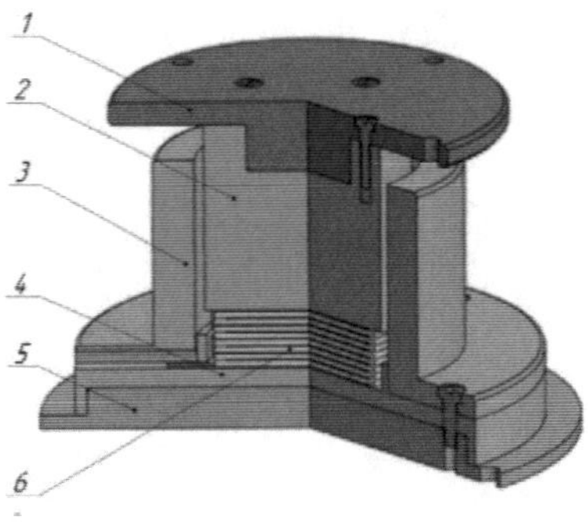

IOP Conf. Series: Materials Science and Engineering **781** (2020) 012007 doi:10.1088/1757-899X/781/1/012007

1– top flange; 2 – rod; 3 – housing; 4 – bottom flange; 5 – disk; 6 – DE

Figure 3. The design of DE-based precise movement mechanism

The DE 6 is attached on the bottom flange, which is a fixed base 4, the rod 2 is mounted on top. The housing 3 is mounted from the top so that the contact pads of the electrodes extend through the grooves for passing the wires out. After aligning all the holes, the housing is attached to a fixed base using a screw connection. The top flange is attached to the rod 2. The assembly is mounted onto disk 5 with a screw connection. Also, there are holes on disk 5 for mounting the mechanism. A movable object is mounted on the top flange 1.

The screw connection of the movable base allows to disassemble the design and replace the DE samples without deforming either the electrodes or the entire elastomer as a whole.

The resulting design can be used for dielectric elastomers with a diameter of not more than 40 mm and a height of not more than 30 mm. The diameter value is determined by the groove in the bottom flange for positioning the DE, and the height should not exceed 75% of the housing height to avoid tilting the movable base.

6. Methods

The goal of the work was to study the properties of DE with quartz as a filler, varying the following parameters: voltage (from 0 to 2000 V), filler concentration (60% and 80%), the number of DE layers (1, 2, 4, 8). The output parameter in this case was the movement of the top base of the actuator.

To carry out the study, an experimental set-up was developed and assembled (Fig. 4), consisting of a power supply unit 1 and a high-voltage power supply unit 2, due to which the necessary potential difference is supplied to the electrodes. The voltage divider 4 allows monitoring the voltage supplied to the actuator electrodes on the voltmeter 3. Inductive sensor 5 shows the movement of the actuator top base, displayed on the control unit of the sensor 7, where one can select the desired measuring range of the sensor. Data from the sensor control unit is sent to the PC screen using ADC 8 and LabView 9 software.

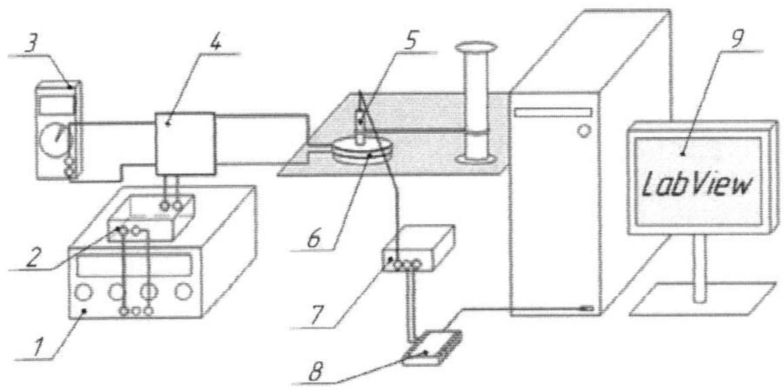

Figure 4. Experimental set-up.

The experiment was carried out according to the following procedure:

1. Setting the required measuring range on the inductive sensor control unit.
2. Setting the required control voltage on the laboratory power supply.
3. Taking the control readings of the output signal of a high-voltage power supply from a multimeter.
4. Recording the value of the current sensor position for 30 seconds, converted using the ADC and the LabView program.

7. Results

During the experiment, voltage was applied to the electrodes in the range from 0 to 2000 V. A series of parallel observations of 150 pieces was taken by an inductive position sensor. Based on the data obtained during the experiment, graphs of the average movement value were plotted depending on the voltage applied to the actuator electrodes, taking into account the number of DE layers (Fig. 5, 6).

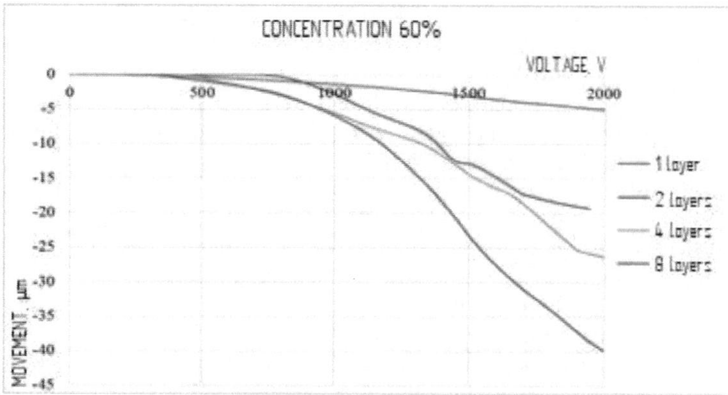

Figure 5. Graph of DE movement with a quartz concentration of 60%

When studying the characteristics of DE with a quartz concentration in the matrix material equal to 60% (Fig. 5), the following maximum movement values were obtained: for 1 layer: -5.01 µm, for 2 layers: -19.25 µm, for 4 layers: -26,32 µm, for 8 layers: -39.81 µm. To study the effect of increasing the filler material concentration in the matrix material, samples with a quartz concentration of 80% were fabricated and studied in the actuator in the same way (Fig. 5).

.

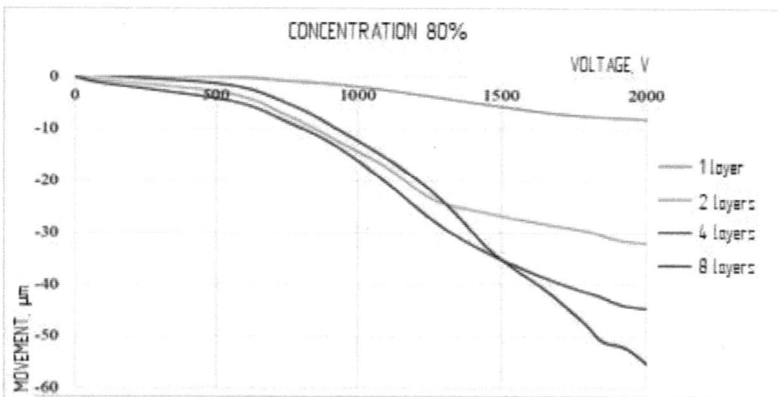

Figure 6. Graph of DE movement with a quartz concentration of 80%

An increase in the quartz concentration led to an increase in the maximum movement value by about 1.6 times: -8.08 µm, -30.55 µm, -44.12 µm, -54.86 µm for one, two, four and eight layers, respectively.

The plotted dependences prove that increasing the number of DE layers allows to increase the movement range of devices based on it, or to reduce the operating voltage required to achieve a certain movement value.

When studying the characteristics of the actuator based on similar DE samples with barium titanate – the "classical" DE filler, the following results were obtained: for one layer -15.39 µm, for two layers -35.35 µm, for four layers -39.11 µm, for eight layers -68.24 microns.

Comparison of the results obtained for DEs based on barium titanate and quartz, we can conclude that quartz can be considered a cheaper analogue of barium titanate for the manufacturing of DEs. Thus, using quartz as a filler, it is possible to achieve a 15 µm worse result on 8 DE layers, but at the same time reduce manufacturing costs by several times.

8.Conclusion

Despite the current state of traditional electronic technologies, there is increasing demand for movement mechanisms that work on new principles. This study showed that dielectric elastomers, depending on their composition and design, can exhibit a wide range of different properties, the use of which in various fields is very promising, in particular – in creation of actuators of high accuracy and efficiency.

It was found that there is a dependence of the actuator top base movement on the number of DE layers, the filler material concentration, and the voltage at the electrodes: an increase in any of the parameters leads to an increase in the output parameter. It is established that for DE manufacturing quartz can be considered a cheaper analogue of barium titanate.

The plotted dependences prove that an increase in the number of DE layers makes it possible to increase the movement range of devices based on it, or to reduce the operating voltage necessary to achieve a certain movement value; the same can be said about increasing the concentration of filler in the matrix.

9. Acknowledgments

The authors express their thanks to the MT11 and MT8 subdepartments of BMSTU for the assistance in organizing and conducting the DE manufacturing process; to Gennady Vladimirovich Stepanov, State Scientific Center of the Russian Federation JSC GNIIHTEOS, for assistance in the technology development and in conducting trial manufacturing processes.

References

[1] Samatham R, Kim K J, Dogruer D, Choi H R, Konyo M, Madden J D and Yim W 2007 Active polymers: an overview. *Electroactive polym. for robotic app.* (London: Springer) pp 1-36

[2] Bauer S 2014 *Dielectric Elastomers. Encyclopedia of Polymeric Nanomaterials* pp 1-9

[3] Bar-Cohen Y 2004 *Electroactive polymer (EAP) actuators as artificial muscles: reality, potential, and challenges* (Bellingham, WA: SPIE press) **136** 1-765

[4] Lochmatter P 2007 *Development of a shell-like electroactive polymer (EAP) actuator* (ETH Zurich).

[5] Brochu P, Pei Q 2012 Dielectric Elastomers for actuators and artificial muscles. *Electroactivity in Polym. Mat.* (Boston, MA: Springer), pp. 1-56

[6] Smela E, Inganäs O, Lundström I 1995 Controlled folding of micrometer-size structures *Science* **268** 1735

[7] Jung M Y, Chuc N H, Kim J W, Koo I M, Jung K M, Lee Y K and Koo J C Mar. 2006 Fabrication and characterization of linear motion dielectric elastomer actuators *Smart Struct. and Mat. 2006: Electroactive Polymer Actuators and Devices (EAPAD)* 6168:616824 International Society for Optics and Photonics

[8] Otero T F, Angulo E, Rodriguez J and Santamaria C 1992 Electrochemomechanical properties from a bilayer: polypyrrole/non-conducting and flexible material – artificial muscle *J. Electroanalyt. Chem.* **341(1-2)** 369-75.

[9] Wang H and Zhu J Nov. 2008 Implementation and simulation of a cone dielectric elastomer actuator *Optomechatronic Tech. 2008* 7266:726607 International Society for Optics and Photonics

[10] Fink J K 2012 *Polymeric sensors and actuators* (John Wiley & Sons) Vol. 1

[11] Carpi F, Chiarelli P, Mazzoldi A and De Rossi D 2003 Electromechanical characterisation of dielectric elastomer planar actuators: comparative evaluation of different electrode materials and different counterloads *Sens. and Act.* A: Phys. **107(1)** 85-95

[12] Brochu P and Pei Q 2010 Advances in dielectric elastomers for actuators and artificial muscles *Macromolecular rapid comm.* **31(1)** 10-36

[13] Tsibizova T Y, Guzeva T A 2016 Automatic control systems for technological processes of curing products made of polymeric composites *Polym. Sci. Ser.* D **9(1)** 22-6

[14] Oohira K 2010 *U.S. Patent No. 7,678,853* (Washington, DC: U.S. Patent and Trademark Office)

[15] Al-Ibadi M 2015 Experimental study of a dielectric elastomer

[16] Rasmussen L (ed) 2012 *Electroactivity in polymeric materials* (Springer Science & Business Media)

[17] Behboodi A and Lee S C K June 2019 Benchmarking of a Commercially Available Stacked Dielectric Elastomer as an Alternative Actuator for Rehabilitation Robotic Exoskeletons *Proc. 2019 IEEE 16th Int. Conf. on Rehabilitation Robotics (ICORR)* pp 499-505

[18] Ma W and Cross L E 2004 An experimental investigation of electromechanical response in a dielectric acrylic elastomer *Appl. Phys.* A **78(8)** 1201-4

[19] Rosset S, Niklaus M, Dubois P and Shea H R 2008 Mechanical characterization of a dielectric elastomer microactuator with ion-implanted electrodes *Sens. and Actuators* A: Phys. **144(1)** 185-93

[20] Pagano C, Malosio M and Fassi I Feb. 2010 Monodirectional Positioning Using Dielectric Elastomers *Intl. Precision Assembly Seminar* (Berlin, Heidelberg: Springer) pp. 180-7

[21] Prahlad H, Pelrine R, Kornbluh R, von Guggenberg P, Chhokar S, Eckerle J and Bonwit, N May 2005 Programmable surface deformation: thickness-mode electroactive polymer actuators and their applications *Smart Struct. and Mat. 2005: Electroactive Polymer Actuators and Devices (EAPAD)* 5759 pp 102-13 International Society for Optics and Photonics

[22] Zarubin V S, Zimin V N and Kuvyrkin G N 2019 Temperature State of a Hollow Cylinder Made of a Polymer Dielectric with Temperature-Dependent Characteristics *J. Appl. Mech. and Tech. Phys.* **60(1)** 59-67

[23] Yoo I S, Reitelshöfer S, Landgraf M and Franke J 2015 Artificial Muscles, Made of Dielectric Elastomer Actuators-A Promising Solution for Inherently Compliant Future Robots *Soft Robotics* (Berlin, Heidelberg: Springer) pp 33-41

[24] Laryushkin P A, Erastova K G, Filippov G S and Kheylo S V 2019 Calculation of Delta-Type Mechanisms with Linear Actuators and Different Numbers of Degrees of Freedom *J. Mach. Manuf. and Rel.* **48(3)** 204-10

[25] Mikhailov V P, Bazinenkov A M, Dolinin, P A and Stepanov G V 2018 Dynamic Modeling of an Active Damper *Rus. Eng. Research* **38(6)** 434-7

IOP Conf. Series: Materials Science and Engineering **781** (2020) 012008 doi:10.1088/1757-899X/781/1/012008

Gas Analysis of Magnetorheological Elastomer During Vacuum Degassing

A M Bazinenkov[1,], I A Efimov[1], A P Rotar[1], D A Ivanova[1]

[1]BMSTU, Moscow, Russia

ambazinenkov@bmstu.ru, dj3dmark@yandex.ru, anastasia.rotari15@gmail.com, Ivanova_D_A@bk.ru

Abstract. Research conducted in a vacuum requires high accuracy, but multiple vibrational and shock disturbing effects adversely affect the results. To increase the accuracy, it is possible to install a vibration isolation system in a vacuum chamber. One of the promising areas of active vibration isolation is the use of intelligent materials in them, including magnetorheological elastomer (MRE). Since most studies need to be carried out in a high and ultra-high vacuum environment, it is necessary to know the degree of influence of the MRE gas release on the quality of the operation environment. Thus, the aim of the work was to analyze the spectrum of released gases during degassing of the polymer (MRE) in vacuum at room and elevated temperatures. It was found that increased gas evolution from the studied polymer sample with increasing temperature does not allow to achieve high vacuum ($1{,}7 \cdot 10^{-3} Torr$). We also identified that gases with the highest partial pressures are part of the atmospheric air; there are other gases not analyzed that significantly affect the total pressure (have a pressure of the order of $10^{-3} - 10^{-4}\, Torr$).

1. Introduction

Among intellectual ("smart") materials, magnetorheological elastomers (MRE) are one of the promising groups of composites. They represent a class of materials with controlled rheological properties i.e. their viscosity, elasticity and plasticity vary depending on the external magnetic field applied to them [1].

The first studies of magnetorheological elastomer were conducted in the 1950s by Jacob Ryabinov in the USA. The composition of MRE is not standardized; various combinations of polymer components are possible. The main elements of MRE is a polymer matrix and filler particles. Experiments on the manufacture of magnetorheological elastomer of various compositions are presented in [2,3,4]. The options of possible filler particles (e.g., FeNdB, Fecarb), as well as their concentrations are shown in these works. The analysis of the compositions allowed us to determine the main components for the manufactured samples. The most commonly used polymers are based on carbonyl iron particles. Particle material has an elongated shape, which contributes to the greatest deformation and rapid magnetization reversal, the optimal concentration of filler particles is 30%.

A study of the mechanical properties of a magnetorheological elastomer is given in [5,6,7,8]. Experimental stands with the possibility of studying deformation under the action of a directed magnetic field are considered. Examples of polymer loading are given to study the possibility of deformation under static and cyclic loads. The magnetic, static and dynamic properties of experimental samples are studied. MRE is used in systems of vibrational protection, in the works [9,10,11,12] the most common constructions of polymer application are given: sensors; converters; automotive bushings and engine mounts; shock protection; self-adjusting vibration absorber.

At the Department of MT11 (BMSTU), an analogue of modern active vibration isolation systems based on the MRE is being developed [13,14]. The modes of operation of this platform and the change in the characteristics of the damper based on the MRE are considered - stability, time constant, positioning error, transmission coefficient of vibration displacement amplitude [15,16,17]. This platform can be applied in a vacuum to protect the objects against vibration in order to improve the accuracy of research and improve the technological capabilities of equipment. To date, the gas evolution of the polymer, including during its degassing, has not been sufficiently studied. The gas evolution of a small sample of a magnetorheological elastomer has previously been considered; experiments showed the possibility of using the polymer in a vacuum [18,19,20].

Thus, the aim of this work was to study the gas evolution of magnetorheological elastomer samples, the sizes of which differ significantly from previously studied. It was necessary to prepare the vacuum pumping station for research, to develop the design of the test chamber, to carry out two test cycles with different warming temperatures, accompanied by heating the chamber, holding at a given temperature and cooling.

2. Methods

MRE gas evolution experiments were carried out at a high-vacuum pumping station with a bell chamber. To conduct the experiment, a chamber was designed with the possibility of placing specimens for tension and compression in it simultaneously, as well as with a volumetric pumping volume [21]. Partial gas pressures were recorded using an Extor 200 quadrupole mass spectrometer.

Preliminarily, on the basis of the laboratory of SSC JSC "GNIIHTEOS", tensile specimens were manufactured in accordance with GOST R 56785-2015 (Polymer composites. Methods of tensile testing of flat specimens) and compression in accordance with GOST 33519-2015 (Polymer composites. Test methods for compression about normal, high or low temperatures) [22].

3.Preparation of Equipment for Conducting Degassing Experiments

Tests to study the gas evolution of the magnetorelastic elastomer were carried out on the basis of the laboratory of NPK Pluton JSC, at a high-vacuum pumping station with a bell chamber (Fig. 1). Pumping is carried out in two stages: fore-vacuum pumping - rotary vane two-stage monoblock pump 2NVR-5DM, high vacuum pumping – magnetic discharge vacuum diode cooled pump - NDMO-01-1. During the operation of the post, the pressure was displayed on the screen of the control unit according to the pressure gauges (wide-range vacuum gauge - SS - 10; thermistor - PMT - 6 - 3M - 1). An Extorr 200 quadrupole mass spectrometer connected to a computer recorded the total pressure in the pumped volume and the partial pressures of the gases specified for the study.

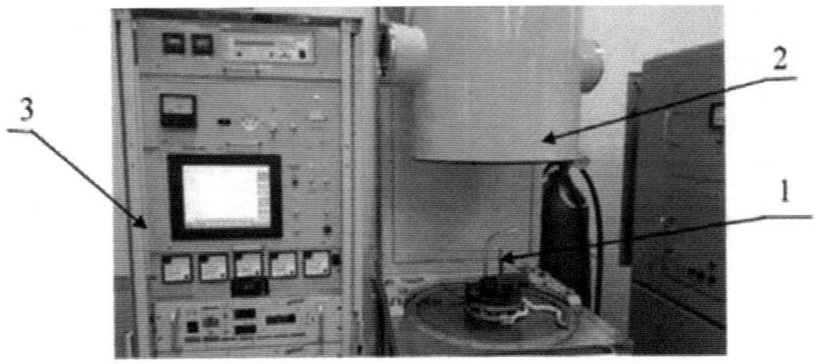

Figure 1. General view of the high vacuum pumping station (1-mount of a tubulation; 2 – bell-shaped vacuum chamber with heaters; 3 – control unit)

For testing, samples were selected for tensile and compression, the placement of which in the tubulation on this installation is impossible. For experiments, as well as for the order of NPK Pluton JSC, a vacuum chamber was designed. Since the installation is a production one, the design change was to be minimal and reversible, so it was necessary to design the camera with fastening on a copper tubulation.

To implement the connection of a copper tubulation with a steel chamber, a circuit similar to the connection with a copper seal of the ConFlat type was proposed (CF) (fig. 2). The connection was established in accordance with GOST 26526 - 85 on vacuum equipment. The upper flange of this connection has CF type teeth, which, when connected to the lower flange, deform the copper plate pre-connected to the copper tubulation, thereby ensuring sealing of this connection.

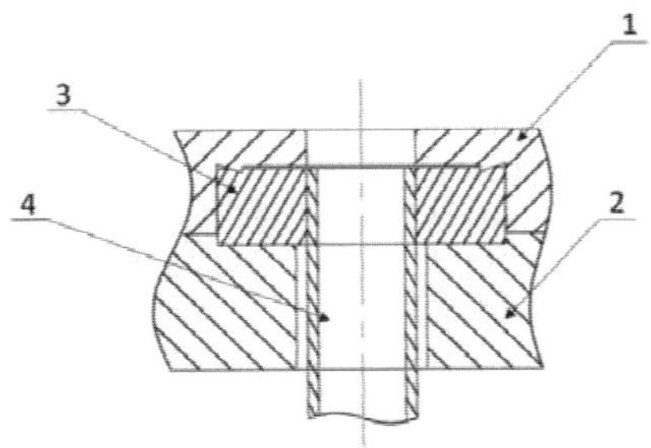

Figure 2. Coupling a copper tubulation with a steel flange (1 – upper flange; 2 – lower flange; 3 – copper plug; 4 – copper tubulation)

To place the samples in the chamber, two hooks were provided that fixed the samples in a predetermined position during pumping (fig. 3). After placing the samples and sealing the upper flange of the chamber (CF copper connection), the design was placed in the plug socket of the installation (fig. 4).

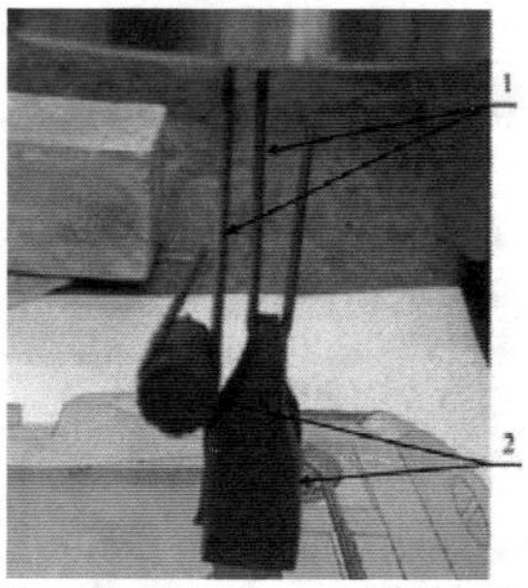

Figure 3. Placement of samples in the chamber (1 – holders; 2 – samples)

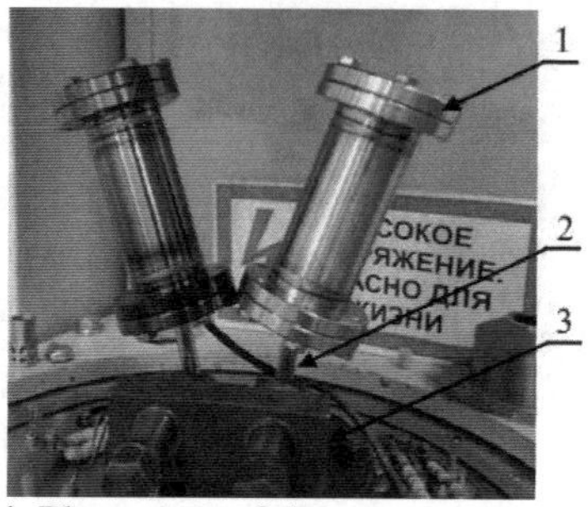

Figure 4. Placement of the camera at the pumping station (1 – chamber; 2 – tubulation; 3 – tubulation holder)

4.Degassing of Magnetorheological Elastomer

One tensile specimen and one compression specimen were placed in the chamber simultaneously (fig. 3). After that, the chamber was pumped out, when the pressure on the magnetic discharge pump of about 10^{-7} 10-7 Torr was reached, the heating process began. Heating continued for 15 minutes, after which exposure was carried out at a given temperature for 30 minutes and uncontrolled cooling of the chamber in air [23,24].

Two consecutive experiments were carried out with heating up to 100^0C and up to 200^0C for different pairs of samples (one for compression and one for tension). A mass spectrometer recorded the partial pressures of the 13 most common gases with atomic masses: 1 u (H); 2 u (H_2); 12 u (C); 14 u (N); 15 u (CH_3); 16 u ($O + CH_4$); 17 u (OH); 18 u (H_2O); 24 u (C_2); 28 u ($N_2 + CO$); 32 u (O_2); 40 u (Ar); 44 u (CO_2).

5. The results

According to the pressure gauges, graphs of the dependence of the total pressure on time in the chamber during pumping were constructed (fig. 5 and fig. 6). The graphs show that the pressure at the beginning of the process (at the start of heating) at different temperatures is similar. Upon reaching the holding temperature, a sharp increase in pressure is observed, which indicates the intensification of gas evolution from pores and from the surface of MRE samples. At the beginning of exposure, the pressure increase is greatest, after 20 minutes the gas evolution rate drops, in Fig. 6 the pressure drop begins. Comparing the pressure during degassing at different temperatures, it is noticeable that at a heating temperature of 200^0C the total pressure in the system reaches about 10^{-3} Torr. -vacuum at this pressure can no longer be considered high.

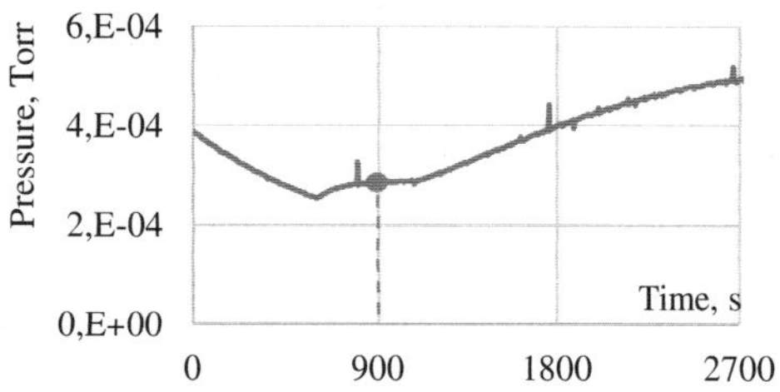

Figure 5. Chamber pressure during warming up to 100 °C and shutter speed

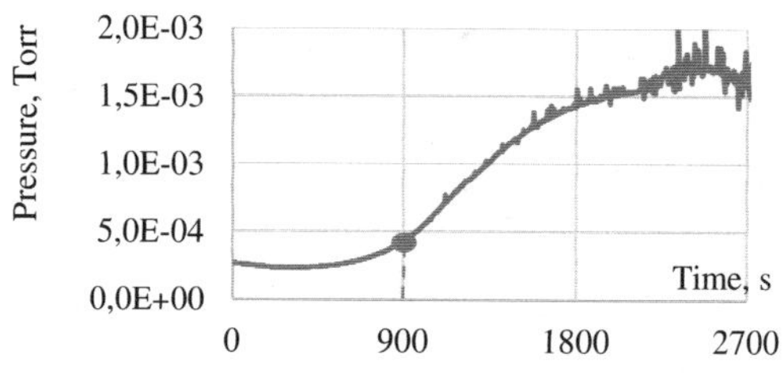

Figure 6. The pressure in the chamber during heating to 200 °C and shutter speed

To analyze the gases that have the greatest influence on the total pressure, graphs of the dependence of pressure on time for each of the 13 studied gases were constructed, all the pressures are of order $10^{-6}...10^{-7}$ Torr (fig. 7 and fig. 8). According to the partial pressure graphs, similar to the total graphs, it is clear that before the start of the degassing process, the pressures practically do not change. When heated up to 100^0C, after the start of the process of degassing, the gas pressure decreases, there are gas peaks indicating gas release from the pores of the MRE. When heated up to 200^0C, a large number of gas peaks do not allow one to judge the degree of pressure reduction by the end of the degassing process. This behavior indicates significant gases in the pores of the samples and other impurities, which are most intensely released with increasing temperature.

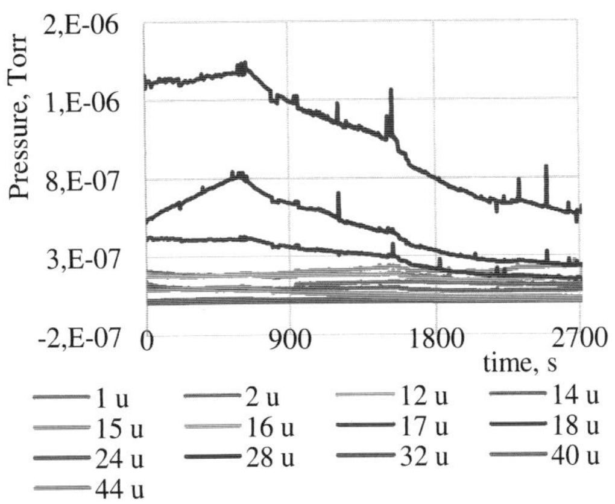

Figure 7. Graphs of partial pressures of gases during heating to 100 °C and exposure

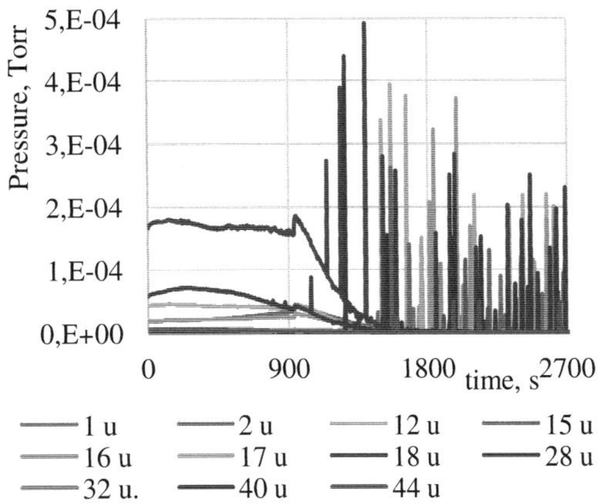

Figure 8. Graphs of partial pressures of gases during heating to 200 °C and exposure

The sum of the partial pressures of the studied gases at a certain point in time differs from the total pressure, which indicates the presence of gases unaccounted for in the study, which have the greatest influence on the studies [25].

After carrying out the experiments and extracting samples from the experimental chamber, oily spots were found on its internal surfaces, and the samples became more dense. Excessive impurities that negatively affected the mass spectrometer readings were probably removed from the material during the degassing process.

6.Conclusion

The designed chamber with tubulation pumping allows to carry out tests at a laboratory pumping station without changing its design. CF copper connection allows high vacuum testing.

Based on the data obtained, it is impossible to unambiguously determine the possibility of using the polymer in a vacuum. With increasing temperature, the flow of gas from the MRE increases, the total pressure in the chamber does not reach the high vacuum level.

Most of the evolved gases are part of atmospheric air and can be removed from the material with longer degassing. In the composition of the evolved gases there are specific ones that have the greatest influence on the degree of vacuum. In the future, it is planned to carry out similar experiments with an extended range of studied gases and to analyze the components removed during pumping.

To conduct further experiments, recommendations will be made on the composition of the polymer, in order to reduce impurity components. It is supposed to conduct experiments with longer exposure and heating to higher temperatures.

7. Acknowledgement
The authors thank Gennady Vladimirovich Stepanov and SSC RF "GNIIHTEOS" JSC for the assistance provided in the manufacture of experimental samples; Ivan Faritovich Khanbekov and NPK Pluton JSC for the assistance provided during the degassing of samples with increasing temperature in vacuum, as well as to Andrei Nikolaevich Rednikin and the RK-5 Department BMSTU for help with tensile and compression tests of MR elastomer.

This work was supported by the Ministry of Education and Science of the Russian Federation, the basic part of State assignment No. 9.8503.2017 / BP in the field of scientific activity.

References

[1.] Boczkowska, A., & Awietjan, S. (2012). Microstructure and properties of magnetorheological elastomers. *Advanced Elastomers-Technology, Properties and Applications*, 147-180.

[2.] Li, Y., Li, J., Li, W., & Du, H. (2014). A state-of-the-art review on magnetorheological elastomer devices. *Smart materials and structures*, *23*(12), 123001.

[3.] Iacobescu, G. E., Balasoiu, M., & Bica, I. (2013). Investigation of surface properties of magnetorheological elastomers by atomic force microscopy. *Journal of superconductivity and novel magnetism*, *26*(4), 785-792.

[4.] Ruddy, C., Ahearne, E., & Byrne, G. (2012). A review of magnetorheological elastomers: properties and applications. *Advanced Manufacturing Science (AMS) Research. http://www. ucd. ie/mecheng/ams/news_items/Cillian% 20Ruddy. pdf Accessed, 20.*

[5.] Boczkowska, A., & Awietjan, S. (2012). Microstructure and properties of magnetorheological elastomers. *Advanced Elastomers-Technology, Properties and Applications*, 147-180.

[6.] Li, J. F., & Gong, X. L. (2008). Dynamic damping property of magnetorheological elastomer. *Journal of Central South University of Technology*, *15*(1), 261-265.

[7.] Alias, N. F., Muthalif, A. G., Arpan, K. A., & Nordin, N. D. (2018). Experimental Investigation of Static Properties of Magnetorheological Elastomer. *Iranian Journal of Science and Technology, Transactions of Mechanical Engineering*, *42*(2), 185-197.

[8.] Wan, Y., Xiong, Y., & Zhang, S. (2018). Temperature dependent dynamic mechanical properties of Magnetorheological elastomers: Experiment and modeling. *Composite Structures*, *202*, 768-773.

[9.] Ahamed, R., Choi, S. B., & Ferdaus, M. M. (2018). A state of art on magneto-rheological materials and their potential applications. *Journal of Intelligent Material Systems and Structures*, *29*(10), 2051-2095.

[10.]Eem, S. H., Koo, J. H., & Jung, H. J. (2019). Feasibility study of an adaptive mount system based on magnetorheological elastomer using real-time hybrid simulation. *Journal of Intelligent Material Systems and Structures*, *30*(5), 701-707.

[11.]Li, W. H., Zhang, X. Z., & Du, H. (2013). Magnetorheological elastomers and their applications. In *Advances in Elastomers I* (pp. 357-374). Springer, Berlin, Heidelberg.

[12.]Ruddy, C., Ahearne, E., & Byrne, G. (2012). A review of magnetorheological elastomers: properties and applications. *Advanced Manufacturing Science (AMS) Research. http://www. ucd. ie/mecheng/ams/news_items/Cillian% 20Ruddy. pdf Accessed, 20.*

[13.]Bazinenkov, A. M., & Mikhailov, V. P. (2015). Active and semi active vibration isolation systems based on magnetorheological materials. *Procedia Engineering*, *106*, 170-174.

[14.]Bazinenkov, A.M., Makeev, I.V., Rotar, A.P., Ivanova, D.A., & Mikhailov, V.P. (2018). Management of the frequency response of the active vibration isolation platform based on magnetorheological elastomers. *Dynamics and vibroacoustics*, *4*(1), 36-40.

[15.]Mikhailov, V. P., Aung, T. L., Kopylov, A. A., & Tovmachenko, D. K. Platform for active vibration isolation of the precision equipment based on magnetorheological elastomers. In *2018 IEEE*

Conference of Russian Young Researchers in Electrical and Electronic Engineering (EIConRus) (pp. 1992-1995). IEEE.

[16.]Mikhailov, V. P., Bazinenkov, A. M., Dolinin, P. A., & Stepanov, G. V. (2018). Dynamic Modeling of an Active Damper. *Russian Engineering Research, 38*(6), 434-437.

[17.]Mikhailov, V. P., Bazinenkov, A. M., Dolinin, P. A., & Stepanov, G. V. (2018). Research on the Dynamic Characteristics of a Controlled Magnetorheological Elastometer Damper. *Instruments and Experimental Techniques, 61*(3), 427-432.

[18.]Ivanova D.A., Rotar A.P. (2018). Gas evolution of magnetorheological elastomer in vacuum. Proceedings of the VIII International Youth Scientific Conference , 92-96.

[19.]Ivanova D.A., Rotar A.P., Bazinenkov A.M. (2019). Study of gas evolution of magnetorheological elastomer during heating in vacuum. Materials of the Ninth Russian Student Scientific and Technical Conference, 65-66.

[20.]Bazinenkov, A. M., Makeev, I. V., Rotari, A. P., & Ivanova, D. A. (2018, July). Estimation of the position error of the magnetorheological elastomers active vibration control platform for precision vacuum equipment. In *IOP Conference Series: Materials Science and Engineering* (Vol. 387, No. 1, p. 012006). IOP Publishing.

[21.]Kalinkin, D. A., Belova, O. V., & Andreev, R. O. (2018, August). Investigation of gas permeability of fibrous composite material in a vacuum. In *AIP Conference Proceedings* (Vol. 2007, No. 1, p. 030034). AIP Publishing.

[22.]Mousokhranov, M. V., Kalmykov, V. V., & Logutenkova, E. V. (2019, March). The influence of technological parameters on physical and mechanical properties of surfaces. In *IOP Conference Series: Materials Science and Engineering* (Vol. 483, No. 1, p. 012054). IOP Publishing.

[23.]Vshivkov, A. Y., Delkov, A. V., Kishkin, A. A., & Lavrov, N. A. (2018). General Concept of Thermal Vacuum Tests of Space-Appliance Elements at Helium Temperatures. *Chemical and Petroleum Engineering, 54*(5-6), 327-334.

[24.]Borodulin, A. S., Malysheva, G. V., & Romanova, I. K. (2015). Optimization of rheological properties of binders used in vacuum assisted resin transfer molding of fiberglass. *Polymer Science Series D, 8*(4), 300-303.

[25.]Svichkar, E. V., Nikulin, N. K., & Demikhov, K. E. (2018, August). Calculation method of pumping characteristics of high-vacuum system with turbomolecular vacuum pump. In *AIP Conference Proceedings* (Vol. 2007, No. 1, p. 030032). AIP Publishing.

IOP Conf. Series: Materials Science and Engineering **781** (2020) 012009 doi:10.1088/1757-899X/781/1/012009

MoS_2 ultrathin films vacuum deposition and its AFM investigation

A I Belikov, Kyaw Zin Phyo, Do Thi Nhan and Aung Linn

Department of electronic technologies in mechanical engineering, Bauman Moscow State Technical University (BMSTU), Moscow 105005, Russia

E-mail: belikov@bmstu.ru

Abstract. In this article, we present the results of AFM studies on molybdenum disulfide ultrathin films (with a thickness of 3 to 13 nm) deposited on silicon substrates in vacuum by the magnetron sputtering MoS_2 target during 10, 20 and 30 seconds. The film samples structural features were estimated and the deposition time influence on surface roughness, crystals size and its distribution asymmetry, fractal dimension was determined.

1. Introduction

The publication's number growth on the subject of 2D-materials in terms of methods and technological modes of their formation is related with an increase in the prospects interest of using 2D-materials in the field of nanoelectronics, and photonics. In recent years, the direction, connected with the production and use of films, based on molybdenum disulfide (MoS_2), shows a rapid increase publication activity. This trend is characterized by a more than fourfold increase in the number of annual publications for 2017 and 2018 compared with 2013 and more than 16 times higher than this index for 2011, when the first report on the implementation of a field effect transistor based on a monomolecular MoS_2 layer appeared. The modern studies results of monolayer's and ultrathin (a few nanometers thick) MoS_2 films electronic-optical properties indicates that this material is promising as a 2D semiconductor material with high electronic properties. In MoS_2 films, the effective charge-carrier mobility can reach 500–1000 $cm^2 \cdot V^{-1} \cdot c^{-1}$ [1]. Field effect transistor based on MoS_2 channel is endowed with high quality switching characteristics. In such transistors, the currents of open and closed states achieved ratio of 10^8 and the bandgap energy can be adjusted by changing the film thickness [3] (the bandgap energy varies from 1.2 eV for bulk material to 1.8 eV in the case of a monomolecular film layer).

Among the currently known formation methods of MoS_2 thin films [4], two groups of methods can be distinguished according to the general implementation principle:

1. "Top-down" – methods that use the natural structural features of MoS_2 crystals (layered structure of a material with weak "Van der Waals" bonds between molecular planes). At the final stage of the implementation, the weakly bound molecular plane is separated from the large crystal and transferred to the substrate. In various methods of this group, physicochemical processes that provide a decrease in the forces of interplanar interaction in a crystal may be used.

2. "Bottom-up" – methods of chemical and physical film synthesis from individual atoms and molecules. It includes CVD methods based on various reactions of the molecules synthesis from precursors with the MoS_2 compound condensation on the substrate surface, or chemical modification methods of the surface with a pre-deposited molybdenum-based film. As well as, it includes PVD methods bottomed with physical sputtering of the MoS_2 target surface with ions of the inert gas and the deposition of an atomized components stream on substrate.

Each group of these methods has specific features that are reflected on the films quality. The methods of the first group are less industrially oriented, but at the same time they provide the formation of high-quality monolayer films on local sections of substrates. The methods of the second group provide a high quality technological environment, controllability and reproducibility of processes; have good scalability of the technology and the ability to obtain uniform films on large substrates.

The magnetron deposition method, which belongs to the second group of methods, is widely used in thin-film technologies. It has been technically developed, ensures high quality and controllability of the deposition processes. The horizons for the commercial use of 2D materials are related to the possibility of reproducibly producing high-quality uniform films on large-sized substrates at relatively low temperatures. It is known that, the structural features, electrical and optical properties of thin films depend on a number of factors in the processes of physical deposition in vacuum. That is why, an urgent task is to assess the capabilities of the magnetron deposition method for the formation of MoS_2 ultrathin films and their structural features identifying. Film growth is accompanied by structural changes, which affected on the morphology and surface roughness, depending on the temperature of the substrate, working pressure, spray power and type of substrate. The released work [5] presents the results, which include studies of the magnetron deposition modes influence on the structural features of MoS_2 thin films.

The aim of this work was to obtain MoS_2 ultrathin films by magnetron deposition at a low discharge power in a short deposition time and to experimentally study their structural features by atomic force microscopy (AFM).

2. Experimental part

In this work, we used a vacuum processing unit that provides pumping of the working chamber to a residual pressure of 10^{-3} Pa. Earlier [6], experiments were carried out to refine the deposition conditions and to study how MoS_2 thin films obtained at various process parameters. Silicon wafers samples with a surface roughness of less than 3 nm were used as substrates. The substrates were preliminary degreased and dried. The substrates were cleaned in alkaline solutions in an ultrasonic bath. The films were deposited by magnetron sputtering of a MoS_2 target in direct current mode. The diameter of the cathode target was 75 mm. In our experiment, the target purity was 99.9%. Argon with 99.9995% purity was used as the working gas. The variable process parameter was the deposition time. Other parameters were constant in all experiments. The working gas pressure was 5×10^{-1} Pa. The discharge power and the magnitude of the current in the magnetron circuit were 22 W and 0.05 A, correspondingly. The substrate temperature was maintained at 200 °C. The distance between the target and the substrate was 100 mm. The deposition modes of the prepared film samples are given in table 1. To assess the films thickness, a mask was applied to the surface. After applying the film, the mask was removed and the height difference was estimated using AFM.

Table.1 Modes of magnetron deposition of MoS_2 films

No sample	Deposition time, s	Substrate temperature, °C	Discharge power, W	Discharge current, A	Argon pressure, Pa
1	10	200	22	0.05	5×10^{-1}
2	20	200	22	0.05	5×10^{-1}
3	30	200	22	0.05	5×10^{-1}

IOP Conf. Series: Materials Science and Engineering **781** (2020) 012009 doi:10.1088/1757-899X/781/1/012009

The structural features of the MoS_2 thin films were evaluated according to surface profile parameters. Scans were prepared at the Solver NEXT AFM manufactured by NT-MDT. Previously, the samples scanning modes were worked out to obtain surface scans with the highest quality. As a result, the contact scanning mode was selected. Scanning was carried out on characteristic areas of the entire film surface that didn't contain visible defects in areas of 0.5×0.5 μm^2 with a surface tapping frequency of 0.5 Hz. Statistical roughness parameters and surface structural parameters were obtained by using IA-P9 and Gwyddion software. The film-substrate step boundary wich was prepared by using a mask, was scanned on an AFM, and the film thickness was measured by the height difference on the step profile. The sizes of crystals (grains), the asymmetry of their distribution, and the fractal dimension were determined. An effective and convenient way to describe the film surface in more detail is fractal image analysis. It is possible to obtain results that allow a more detailed analysis and classification of the film surface morphology and the calculation of the values of its fractal dimension by using fractal analysis of AFM surface images. To determine the morphological fractal dimension, we used the method of counting cubes and the method of triangulation, based on the description of the shape of a three-dimensional surface profile using geometric constants. In order to determine the fractal dimension of the films, computer processing of micro images of the deposited MoS_2 films was carried out in the Gwyddion program.

3. Results and discussion

Figure 1 shows images of the silicon substrate surface (a) and the MoS_2 film formed during 10 s deposition time (b). MoS_2 film slightly follows the relief of the substrate; still, it has a smoother surface with lower roughness. So, the roughness of the substrate and the film are equal, respectively, 1.181 nm and 0.464 nm. These results and images of the surface topography of the samples indicate a noticeable smoothing of the surface after film deposition. The films thicknesses were measured in various cross sections of the "film-substrate" step-interface. The film thickness for sample No.1 equaled 3 nm, for sample No.2 the thickness was 7 nm and for sample No.3 – 13 nm.

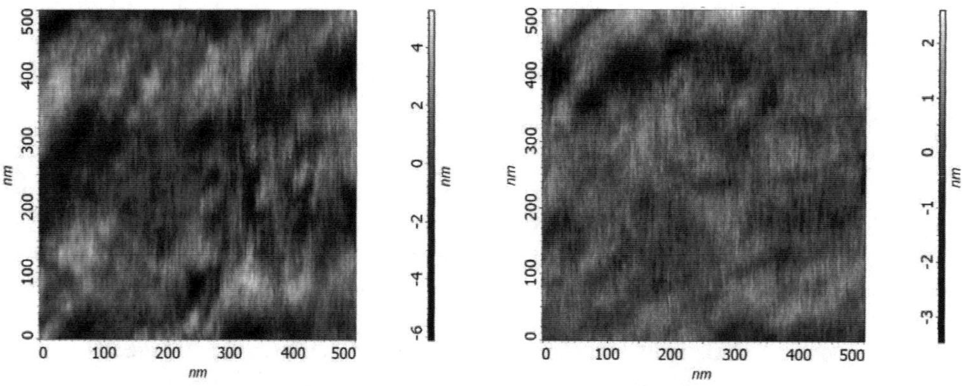

(a) (b)

Figure 1. AFM images of silicon substrate (a) and MoS_2 film sample deposited in 10 seconds (b)

Characteristic features of the samples surface areas with MoS_2 films are shown in Fig. 2. The main parameters of the film surface are presented in Table 2.

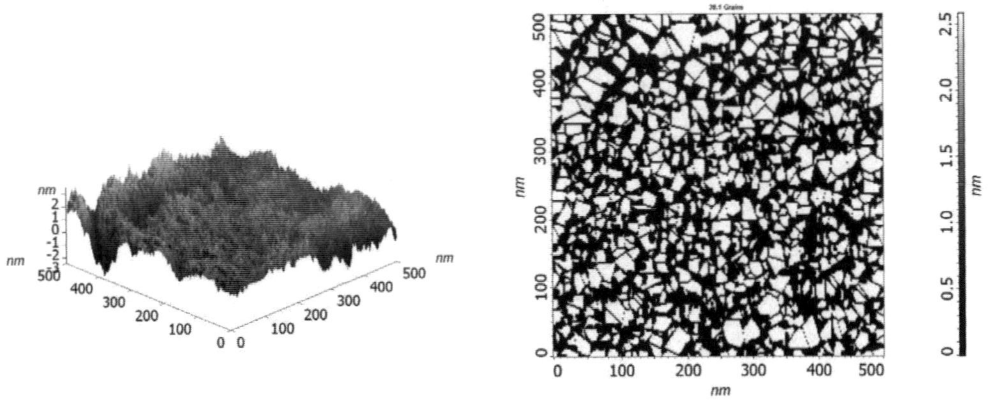

a) Sample No.1, MoS_2 on Si, 10 s.

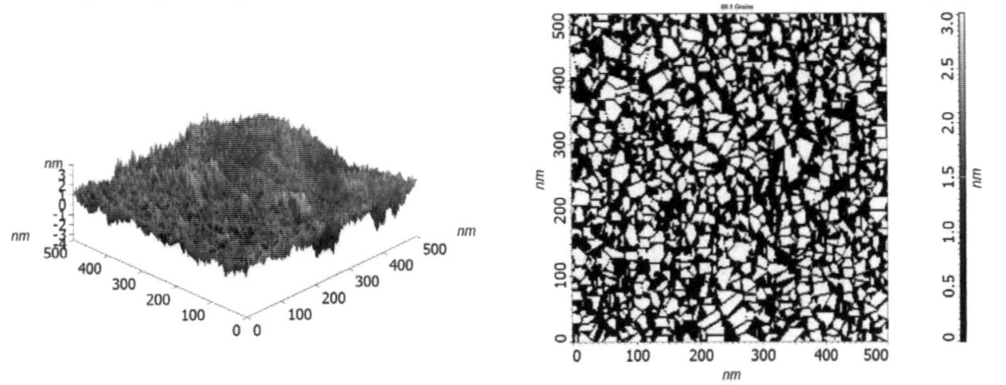

b) Sample No.2, MoS_2 on Si, 20 s.

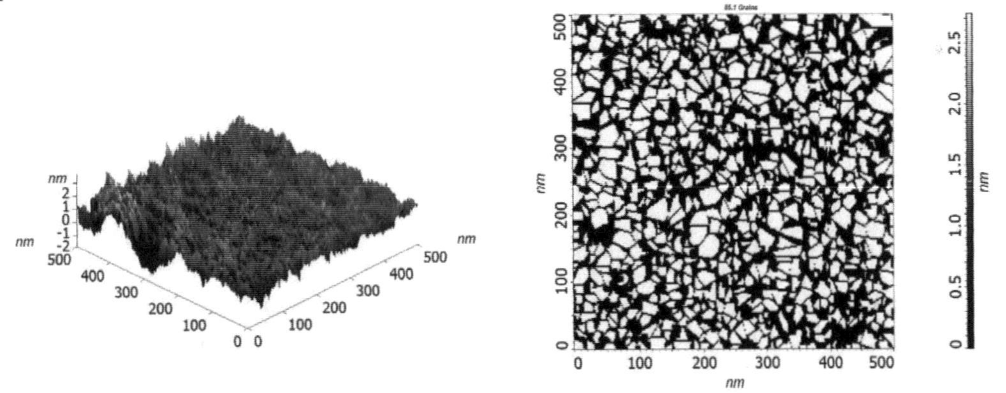

c) Sample No.3, MoS_2 on Si, 30 s.

Figure 2. AFM 3D images and grain distribution on 0.5×0.5 mm^2 surface areas for MoS_2 films deposited over different times: (a) sample No.1 (10 s); (b) sample No.2 (20 s) and (c) sample No.3 (30 s).

As the result of deposition for 10 seconds a continuous film with a roughness of 0.464 nm with an average grain size of 11.805 nm is formed on the surface of the sample No.1. An increase in the application time to 20 seconds (Fig. 2b) leads to an increase in the number of grains, the average grain size decreases slightly (to 11.551 nm), and the roughness and surface area increase. Film grain

IOP Conf. Series: Materials Science and Engineering **781** (2020) 012009 doi:10.1088/1757-899X/781/1/012009

coarsening is observed with a further increase in the deposition time up to 30 seconds (Fig. 2c). Also, the average grain size increases to 12.695 nm, and the roughness and actual surface area decrease.

Table.2 Results of surface parameter's measurements of MoS_2 films

Parameter \ Deposition time,s	10	20	30
Arithmetic mean roughness Ra, nm	0.464	0.521	0.371
Asymmetry coefficient, *Rsk*	-0.393	-0.229	-0.170
Surface area, μm^2	0.257	0.263	0.256
Average grain size, nm	11.805	11.551	12.695

Since the asymmetry coefficient of the structure Rsk differs from zero, we can talk about the uneven crystallites distribution relative to the center of the studied area. Sample No.1 is characterized by the largest asymmetry coefficient, which indicates a significant uneven distribution of crystallites.

The fractal dimension values obtained by using the cubic counting method and the triangulation method for a scanned surface with an area of 0.5×0.5 μm^2 are shown in Table 3. In contrast to the root-mean-square roughness, the fractal dimension of the films surface demonstrates scale invariance in the studied range of AFM image sizes. Table 3 shows us, the values of the fractal dimension calculated by different methods are differ from each other. In addition, the used methods are characterized by different accuracy in calculating the fractal dimension. It is known that the method of counting cubes has a smaller error in determining the value, in contrast to the triangulation method. Some surface smoothing irregularities occurs to the fact that during fractal analysis we use the triangulation method. Also, averaging the results entails less accuracy in the estimation of fractal dimension. Therefore, we should consider the results which were obtained by using the method of calculating cubes.

Table.3 Fractal dimension of MoS_2 films deposited at different times (based on the results of AFM image processing by different methods).

Deposition time, s	Cube counting method	Triangulation method
10	2.406	2.453
20	2.438	2.482
30	2.443	2.493

The values of the fractal dimension obtained by using the method of counting cubes vary in a fairly narrow range from 2.406 to 2.433. The increase in fractal dimension depends on the deposition time increase. As a result, the surface morphology becomes more uniform, causing a gradation of its fractal properties. In the light of the lower roughness and actual surface area obtained for films with deposition times of 10 and 30 s, it is possible to note the presence of a higher planarity of the surfaces of these samples in comparison with sample No.2. This situation may be characteristic of the mechanism of two-dimensional growth of islands according to the type of layer-by-island growth. An increase in surface roughness was observed at the 20 s deposition stage. It may be caused with new islands grows out at planar surfaces previously formed. With further deposition (up to 30 seconds), two-dimensional growth of the islands leads to a decrease in roughness and an increase in planarity of the film surface.

4. Conclusions

The study's results indicate a significant (more than 2.5 times) decrease in the surface roughness of silicon substrates after the MoS_2 films deposition. The most probable, this may be due to the predominance of the preferential incorporation processes of MoS_2 molecules into the nodes formed by steps and kinks of the surface and the presence of a mechanism of two-dimensional islands growth and steps at the initial stage of film growth.

The minimum thickness of the obtained MoS_2 films was 3 nm and it was achieved by deposition of a particles stream sputtered from the target for 10 seconds. The thickness of the films obtained by deposition for 30 seconds was 13 nm. As a result of our work, no significant effect of the deposition time on the roughness and grain size of the films were revealed. However, with an increase in the deposition time, a steady decrease in the asymmetry coefficient and a slight increase in the fractal dimension were observed.

References

[1] Baugher B W et al. 2013 *Nano letters***13** 421–6
[2] Duan X et al. 2015 *Chem. Soc. Rev***44** 8859
[3] Mak K F, Lee C, Hone J, Shan J, Heinz T F 2010 *Phys Rev Lett.***105** 136805
[4] Belikov A I, Kyaw Zin Phyo 2016 *Fund.Probl.of Radio-Electr.Instr.Making***16** 12–16
[5] Belikov A I, Do Thi Nhan, Kalinin V N 2018 *IOP:Mat.Sci.and Eng.***387** 012007
[6] Belikov A I, Kyaw Zin Phyo, Kalinin V N, Semochkin A I 2019 *Proc.of SPIE***11022** 110221F

IOP Conf. Series: Materials Science and Engineering **781** (2020) 012010 doi:10.1088/1757-899X/781/1/012010

Development of a calculation method of the main parameters of the multistage turbomolecular pump

U. S. Gordeeva[1 a)], K. E. Demikhov[1 b*)], A. A. Ochkov[1 c*)]

[1] Bauman Moscow State Technical University, Russian Federation, Moscow, 2nd Baumanskaya st., 5.

[a)] Corresponding author: tests.ibmes@gmail.com
[b*)] kdemikhov@yandex.ru
[c*)] aochkov@bmstu.ru

Abstract: We developed a calculation method of the pumping parameters of the molecular stage of the turbomolecular vacuum pump (TMP) operating in the molecular flow region. Also, the authors developed the calculation method of the main parameters of the multistage TMP for the providing the main pumping parameters including ultimate pressure, pumping speed, compression ratio. As a result of calculation, the we got the performance curve of the multistage TMP.

1. Introduction

Turbomolecular vacuum pumps (TMP) are widely used in R&D and industry applications. According with the high requirements for vacuum systems, the problem of their layout to ensure the technological process is relevant. Turbomolecular vacuum pumps have a number of advantages compared to other mechanical vacuum pumps, namely they do not pollute the pumped volume, provide a low ultimate residual pressure without the pumping selectivity [1-4]. Molecular vacuum pumps are currently rarely used separately, due to the relatively low pumping speed. To increase the pumping efficiency, the molecular stage is combined with the turbomolecular stage of TMP [5-8]. There are three most common design groups of flow channels: TMP and a cylindrical stage with annular channels on the rotor (Gaede model Fig. 1a), TMP and a cylindrical stage with spiral channels along the rotor surface (Holweck model Fig. 1b), TMP and a disk stage with spiral channel from the outer diameter to the centre of the disk (Siegbahn model Fig. 1c) [9,10]. The review of references showed that in the presence of various structural configuration of multistage flow channels of the TMP, no single method for calculating such structures was revealed, which determines the relevance of the study.

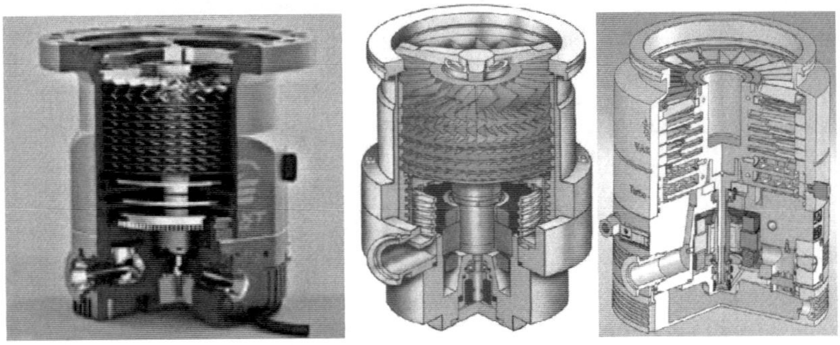

Fig. 1. Multistage turbomolecular pumps:

a – Gaede model; b – Holweck model; c – Siegbahn model

2. Calculation method of the main parameters of the multistage turbomolecular pump

The structure of the flow channel of the considered TMP is determined by rotor with two stages such as turbomolecular and molecular.

In turn turbo stage is consist of the number of blades packages which provides the pumping speed required [11].

The modelling assumptions are the following:

1. The velocity distribution of the molecules is in accordance with the Maxwell distribution law in the inlet and forevacuum sections;
2. Molecular flow regime in the channel;
3. The process of gas molecules transfer is considered quasi-steady-state.

To determine the impeller geometry, theoretical models of the transition of gas molecules through the wheel are used taking into account the interaction with the walls of the blades channel. To determine the coefficient of probability of gas molecules transition through the impeller (K_{max}), it is necessary to determine the permissible peripheral speeds of the wheel (u_2), the arithmetic average velocity of the thermal motion of the molecules of the pumped gas ($v_{aver.}$), and also the geometry of the operating channel [12].

To determine the pumping parameters of the wheel, it is necessary to determine the actual maximum pumping speed of the wheel ($S_{max.act}$), taking into account the obtained geometric parameters. To calculate a pumping curve of the wheels, it is necessary to determine the compression ratio of the wheels for each of the packages (τ_{max}) [13].

$$\tau_{max} = \frac{\int_{\lambda}^{1} \left[r k_{ab1} + \frac{h_1}{H_1} \frac{(1-\lambda\psi)(r-\lambda)}{(1-\lambda)} \right] \int_{0}^{F} \frac{\gamma+\beta_{\gamma}}{\pi F} dx dr}{\int_{\lambda}^{1} \left[r k_{ab1} + \frac{h_1}{H_1} \frac{(1-\lambda\psi)(r-\lambda)}{(1-\lambda)} \right] \int_{0}^{F} \frac{\gamma-\beta_{\gamma}}{\pi F} dx dr} \quad (1)$$

where ψ is blade thickness ratio,
$k_{ab\,1}$ is channel width ratio,
h_1 is blade thickness,
H_1 is width of the operating channel,
λ is relative diameter of the wheel,
r is relative radius,
γ, β_{γ} is probability of transition of gas molecules through the blade channels.

In multistage TMP the molecular drag stages operate in molecular and viscous flows. W. Gaede model is used to obtain the calculations formulas of the molecular drag stage [14,15].

To obtain the main parameters of the stage pumping curve it is necessary to determine maximum pressure drop in the channel $(p_1 - p_2)_{max}$.

$$(p_1 - p_2)_{max} = \frac{6\eta Lu}{h^2} \tag{2}$$

where η is dynamic viscosity,
$L-$¿is channel length,
$u-$¿is rotary rotations per minute,
$h$is channel height.

Next, it is necessary to determine the maximum pumping speed of the channel$(p_1 - p_2)_{max}$.

$$S_{max} = \frac{mRT}{M\,p_2} \tag{3}$$

where m is the maximum gas amount pumped by the channel at equal pressures;
R is universal gas constant;
$T$is temperature of the gas pumped;
M is molecular mass of the gas pumped.

To determine the pumping speed of molecular part, it is necessary to interconnect the geometric, speed and pumping parameters of the molecular part of channel [16,17].

$$x_1 - x_2 = \frac{h^2}{6\eta(u_1+u_2)}(p_1 - p_2) + \left[\frac{h}{\Theta(u_1+u_2)} + \frac{mRTh}{Mb\,3\,\eta(u_1+u_2)^2}\right] * \ln \frac{p_1 - \frac{2mRT}{Mbh(u_1+u_2)}}{p_2 - \frac{2mRT}{Mbh(u_1+u_2)}} \tag{4}$$

where $x_1 - x_2$ is channel length;
$u_1; u_2$is speed of the walls;
Θ is coefficient of external friction;
b is breadth of the channel.

It is also necessary to determine the maximum pressure ratio τ_{max}.

$$A\left(\tau_{max} + \frac{1}{\tau_{max}} - 2\right) + B(\tau_{max} - 1) - C lg\,\tau_{max} + lg^2\tau_{max} = 0 \tag{5}$$

where $A = \frac{(2\pi r h)^3}{Lb(2{,}303h)^2}$;

$$B = \frac{Suh\Theta(2\pi r)^3}{b(2{,}303\,h)^2};$$

$$C = \frac{(u_1+u_2)2\pi r\Theta}{2{,}303h}.$$

To calculate the pumping curve of the pump, the pumping speed of the multistage flow channel is determined S_N [18-20].

$$S_N = S_{раб.} + \frac{Q_{гв.вс.}}{p_1} \tag{6}$$

where $Q_{гв.вс.}$ is gassing in the inlet suction.

3. Results and discussions

The developed calculation method was used to obtain the pumping curve of the multistage TMP. The pumping curve is shown in Fig. 2.

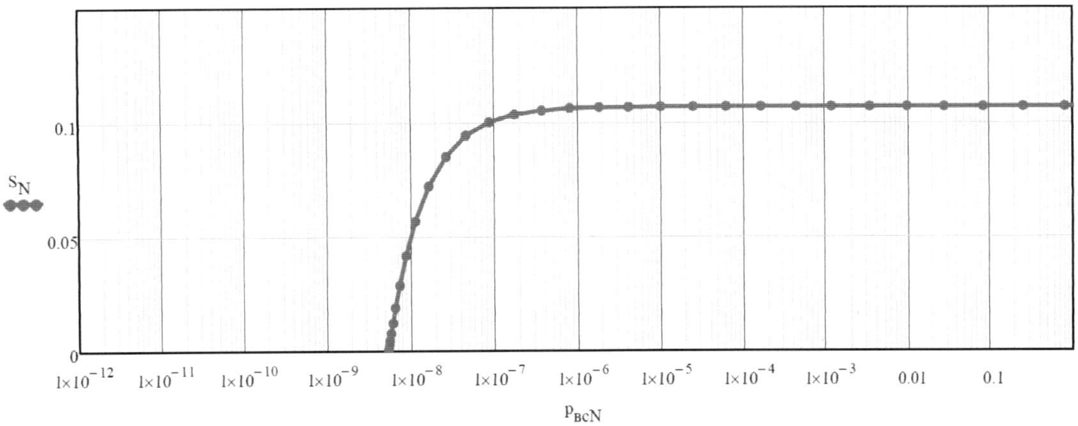

Figure 2. Pumping curve of the multistage TMP

4. Conclusion

Due to the fact that TMP are most often used in industrial and research practice, increasing of the pumping speed and the maximum ultimate pressure are the most important problems. TMPs with multistage flow channel have the greatest potential for the effective implementation of tasks.

The developed calculation method allows determining the main parameters of the TMP with a with multistage flow channel such as: compression ratio of each stage, the pressure drop, the pumping speed of the pump.

As a result of the calculations, the pumping curve of a multistage TMP was obtained.

References

[1] Demikhov, K. E., Nikulin, N. K., Svichkar, E. V. Prospects for the development of combined turbomolecular vacuum pumps. Engineering Journal: Science and Innovation, (5), - 2013, 36-36.

[2] Demikhov KE. Modern trends in the development of high-vacuum mechanical pumps // Herald of BMSTU. Series "Engineering". - 2014. - No. 5 (98).

[3] Frolov, E. S., Avtonomova, I. V., Vasiliev, V. I., Nikulin, N. K., Plastinin, P. I. Mechanical vacuum pumps. Production Edition. Engineering. -1989.

[4] Demikhov KE. The main approaches to the selection of optimality criteria in the design of high-vacuum mechanical pumps // Herald of BMSTU. Series "Engineering". - 2015. - No. 4 (103).

[5] Pugachuk A. S. et al. Development of radial turbines in low-power Gas-turbine engines //Journal of Physics: Conference Series. – IOP Publishing, 2019. – V. 1163. – №. 1. – S. 012022.

[6] Demikhov K. E., Ochkov A. A., Gordeeva U. S. Development of calculation method of main parameters of vacuum liquid transportation system //IOP Conference Series: Materials Science and Engineering. – IOP Publishing, 2018. – V. 387. – №. 1. – S. 012014.

[7] Djuzhev N. A. et al. An experimental study on MEMS-based gas flow sensor for wide range flow measurements //2018 IEEE Sensors Applications Symposium (SAS). – IEEE, 2018. – S. 1-4.

[8] Vasilev D. D., Malevannaya E. I., Moiseev K. M. Vacuum coating system for deposition of superconducting W x Si (1–x) ultrathin films used in single photon detectors //Journal of Physics: Conference Series. – IOP Publishing, 2017. – V. 872. – №. 1. – S. 012027.

[9] Arzymatov B., Deulin E. Solid lubricant mass contact transfer technology usage for vacuum ball bearings longevity increasing //Journal of Physics: Conference Series. – IOP Publishing, 2016. – V. 729. – №. 1. – S. 012027.

[10] Demikhov K. E., Ochkov A. A. Software optimization of the main parameters of turbomolecular vacuum pumps // Engineering Journal: Science and Innovation. - 2013. - No. 5. - P. 17.

[11] Vacuum equipment / K.E. Demikhov, Yu.V. Panfilov, N.K. Nikulin and others / under the total. ed. K.E. Demikhov, Yu.V. Panfilov. M .: Mechanical Engineering, 2009.590 c.

[12] Demikhov K.E., Nikulin N.K. Optimization of high vacuum mechanical pumps. M.: Publishing House of BMSTU, 2010.225 s.

[13] Demikhov K. E. Features of optimization of the flow part of high-vacuum mechanical pumps in a wide pressure range // Herald of BMSTU. Series "Engineering". - 2012. - No. 3.

[14] Demikhov K. E., Ochkov A. A. Universal mathematical model of the process of pumping gas by a molecular vacuum pump // Herald of BMSTU. Series "Engineering". - 2017. - No. 6 (117).

[15] Demikhov K. E., Ochkov A. A. Determination of the effective range of gas pressures on the suction side of a turbomolecular vacuum pump // Herald of BMSTU. Series "Engineering". - 2016. - No. 5 (110).

[16] Demikhov K. E., Ochkov A. A., Polezhaev A. Influence of various parameters of the flow part of a cylindrical molecular vacuum pump on its characteristics // Machines and Installations: Design, Development and Operation. - 2015. - No. 3. - S. 1-8.

[17] Becker W. The turbomolecular pump, its design, operation and theory; calculation of the pumping speed for various gases and their dependence on the forepump //Vacuum. – 1966. – V. 16. – №. 11. – S. 625-632.

[18] Demikhov K. E., Ochkov A. A., Tsakadze G. T. Method for calculating the optimal parameters of a combined molecular vacuum pump // Herald of BMSTU. Series "Engineering". - 2017. - No. 5 (116).

[19] Svichkar E. V., Nikulin N. K., Demikhov K. E. Method for calculating the pumping characteristics of a high-vacuum system with a turbomolecular vacuum pump // Omsk Scientific Herald. Series "Aircraft-rocket and power engineering." - 2018. - T. 2. - No. one.

[20] Demikhov K. E., Ochkov A. A. Determination of the effective range of gas pressures on the suction side of a turbomolecular vacuum pump // Bulletin of Moscow State Technical University. NE Bauman. Series "Engineering". - 2016. - No. 5 (110).

IOP Conf. Series: Materials Science and Engineering **781** (2020) 012011 doi:10.1088/1757-899X/781/1/012011

Investigation of WSi and NbN superconducting single-photon detectors in mid-IR range

A V Antipov[1,2], V A Seleznev[1,2], Yu B Vakhtomin[1,2], P V Morozov[1], D D Vasilev[3], E I Malevannaya[3], K M Moiseev[3] and K Smirnov[1,2,4]

[1] Moscow Pedagogical State University, 119991, Moscow, Russia
[2] LLC Superconducting nanotechnology (Scontel), 119021, Moscow, Russia
[3] Bauman Moscow State Technical University, 105005, Moscow, Russia
[4] National Research University Higher School of Economics, 101000, Moscow, Russia

E-mail: smirnov@scontel.ru, d.d.vasiliev@bmstu.ru, k.moiseev@bmstu.ru

Abstract. Spectral characteristics of WSi and NbN superconducting single-photon detectors with different surface resistance and width of nanowire strips have been investigated in the wavelength range of 1.3-2.5 μm. WSi structures with narrower strips demonstrated better performance for detection of single photons in longer wavelength range. The difference in normalized photon count rate for such structures reaches one order of magnitude higher in comparison with structures based on NbN thin films at 2.5 μm.

1. Introduction

Presently superconducting single-photon nanowire detectors (SNSPD) have found wide application only in those areas where it is necessary to detect single photons in the visible and near-infrared range 0.4-1.5 μm [1-9]. However, when we move to the longer wavelength range 1.5-4 μm, quantum efficiency of such detectors drops significantly [10] making them almost untapped not only in the field of practical applications but also in various scientific fields. In accordance with the model of the "hot spots" [11] the principle of operation single photon detector based on photon absorption by narrow and thin superconducting strip with the formation of the resistive region. If we move towards longer wavelengths, we have to reduce the cross-section of the strip in order to decrease the photon energy required for the destruction of superconductivity within this area [10]. However, this approach has its own limitations due to manufacturing issues in the detector development process. For example, the roughness of the strip edges plays significant role when its width approaches the limits of electron beam lithography. The width deviation of 5% gives a significant reduction in the quantum efficiency by one order of magnitude. In addition, narrowing the width of the strip leads to a reduction of the effective area of the detector [12]. In order to increase the quantum efficiency, we need to increase the filling factor of the meander [13]. However, it leads to an increase in the overall length of the superconducting strips with increasing its kinetic inductance which limits the performance of the detector [14]. The fabrication of such structures encounters technical difficulties due to additional requirements for high quality fabrication of long and narrow superconducting strips. Another disadvantage is the reduction of the critical current in such superconducting structures which leads to the decrease in the amplitude of the recorded signal and signal-noise ratio giving the worse jitter.

Moving towards longer operating wavelength region of single-photon detectors can contribute to the investigation of materials with lower energy gap or with lower superconducting transition temperature. Among such materials we can mark WSi [14] and VN [15]. Amorphous WSi does not require high quality substrate in comparison with detectors based on NbN where the quantum efficiency depends on the crystalline phase of the NbN [16] or defects on the surface of the substrate. Another direction to increase the quantum efficiency of the detectors is the reduction of the thickness of the superconducting film or fabricating films with more disordered structure [17]. This direction does not require high resolution electron beam lithography is therefore more technological.

Therefore, it seems reasonable to investigate the properties of superconducting films based on different materials such as WSi, and NbN, varying both the disorder degree in the films and its geometrical parameters, providing a high quantum efficiency of the detectors with low level of dark counts and jitter.

2. SNSPD fabrication technology and experimental setup

Thin NbN and WSi films were deposited on sapphire and silicon substrates by reactive magnetron sputtering in AJA International Orion-8 [18] and VUP-11M, correspondently [19, 20]. The thickness of such a film was ~7 nm, the surface resistance was 300-500 Ohm/sq. The width was 80-100 nm. The strip has a meander shape and covers a square area of 15x15 μm2 with a filling factor (the ratio of the area occupied by the superconducting meander with respect to the nominal area of the device) as high as 0.5–0.6. The critical temperatures are T_C = 8-10 K for NbN films and T_C = 2.7-3.3 K for WSi films. Superconductive strips were formed in a gas mixture of SF_6 and argon by e-beam lithography and subsequent plasma-chemical etching. Au contact pads were made using the lift-off process and resistive evaporation. Photolithography and subsequent plasma chemical etching were used in order to remove the superconducting film from the substrate surface except the working region and contact areas. The wafer was cut into separate chips by using scriber.

To determine the relative number of counts, we used a standard technique for measuring the input in-fiber radiation power at a particular wavelength and the number of voltage pulses appearing on the detector during absorption of this radiation. Radiation from infrared spectrometer was coupled with the detector trough multimode fiber. The output power from the fiber at different wavelengths was calibrated with InAs infrared detector (Hamamatsu P10090-21). The spectral range of measurements was limited by the wavelength of 2500 nm due to bandwidth of the multimode optical fiber. The samples were cooled down to the temperature of 1.7 K by pumping helium vapors from a thermally-insulated insert. The electrical output signal and the detector bias current were transferred through CuNi coaxial cables and hermetically sealed SMA connectors. It should be noted that films with close surface resistances were selected to compare the spectral characteristics of samples based on different materials.

3. Results and discussion

We have studied the relative spectral quantum efficiency of detectors based on superconducting WSi and NbN films with different surface resistance. Figure 1 shows the relative number of counts vs. wavelength of incident radiation for NbN and WSi structures with different surface resistances and strip width.

The bias current for all samples was set to $0.9Ic$, where Ic is the critical current of superconducting film. We normalized all dependencies at the wavelength of 1300 nm in order to observe more clearly the behavior of relative counts. It can be seen that NbN structure with narrower strip width of 80 nm and lower surface resistance *Rs* 340 Ohm shows dramatic exponential decreasing of counts by two orders of magnitude with increasing of wavelength by almost two times from 1300 nm to 2500 nm. However, NbN structure with almost the same strip width of 84 nm and higher surface resistance *Rs* 487 Ohm exhibits one order of magnitude decrease in the number of counts. Such behavior of counts for two samples with different surface resistance is a reasonable expectation because absorption coefficient should be higher for high resistive superconducting films. WSi structures show weaker dependencies of counts in comparison with the structures based on NbN films. The number of counts for WSi structure with strip width of 100 nm decreases by two times (Figure 1 and 2). However, the structure with strip width of 80 nm exhibit fluctuation in counts within 20%. The narrowest WSi strip gives the highest absorption which stays almost constant within a wavelength range of 1300-2500 nm.

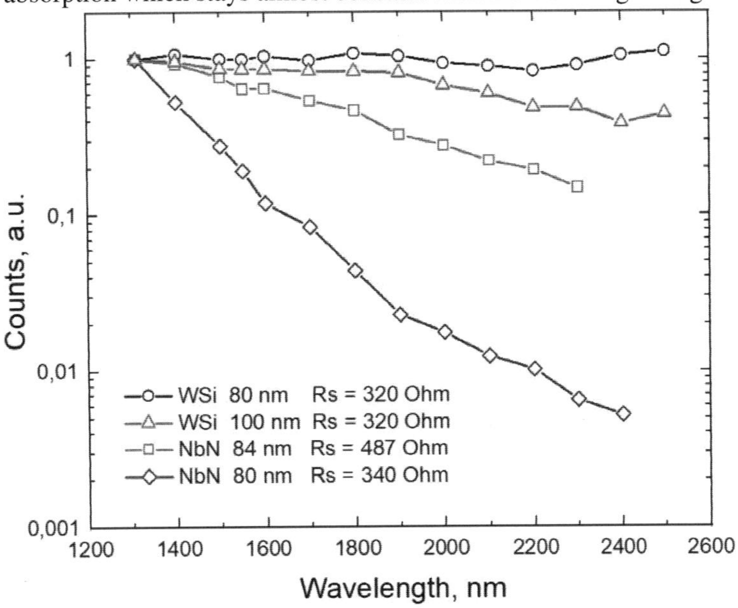

Figure 1. The relative number of counts vs. wavelength of incident radiation for NbN and WSi structures with different surface resistances and strip width. The bias current is set to 0.9*Ic*. All counts are normalized at 1300 nm.

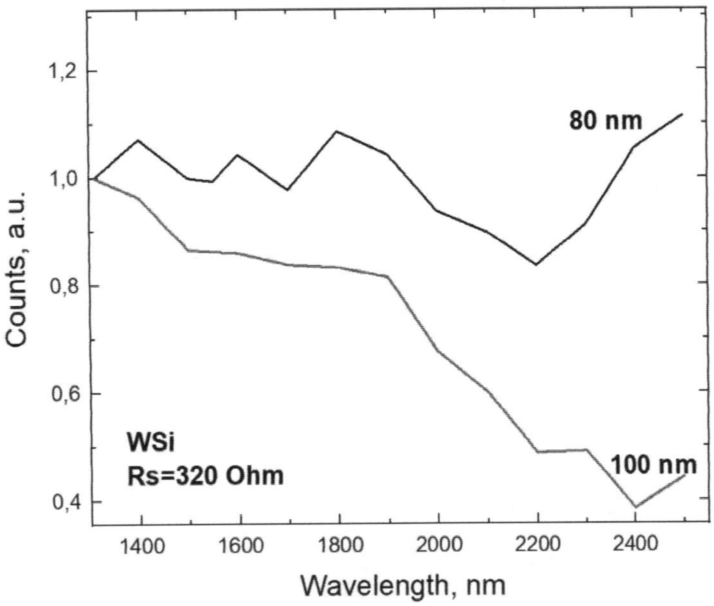

Figure 2. The relative number of counts vs. wavelength of incident radiation for WSi structures with surface resistances *Rs* 320 Ohm and different strip width 80 nm and 100 nm. The bias current is set to 0.9*Ic*. All counts are normalized at 1300 nm.

Thus, the obtained measurements show that we can substantially improve the performance of single photon detectors towards longer operating wavelengths for practical applications by choosing WSi structures with the narrowest strip width and the highest film surface resistance. However, detectors based on WSi thin films have several significant disadvantages in comparison with NbN films. The critical temperature of superconducting WSi thin films are about 3 K, which is lower than the temperature of liquid helium 4.2 K. Also, they have low critical current about Ic 9 μA at 1.7 K. It requires additional special cryogenic equipment to reach the temperature below 4.2 K, and low-noise amplifiers to get appropriate signal-noise ratio.

4. Conclusions

We investigated spectral characteristics of WSi and NbN superconducting single-photon detectors with different surface resistance and width of nanowire strips in the wavelength range of 1.3-2.5 μm. WSi structures with surface resistances Rs 320 Ohm and strip width of 80 nm demonstrated better performance for detection of single photons in longer wavelength range. Superconducting single-photon detectors based on NbN and WSi thin films exhibit significant difference in quantum efficiencies for the long-wavelength region, which reaches one order of magnitude at 2.5 μm. However, the proper operation of superconducting WSi structures requires deeper cooling in comparison with NbN structures due to its low critical temperature.

5. Acknowledgment

The research is supported by the Russian Science Foundation project No. 18-12-00364.

References

[1] Divochiy, F. Marsili, D. Bitauld, A. Gaggero, R. Leoni, F. Mattioli, A. Korneev, V. Seleznev, N. Kaurova, O. Minaeva, G. Goltsman, K. G. Lagoudakis, M. Benkhaoul, F. Levy, and A. Fiore 2008 *Nat. Photonics* **2**, 302

[2] Korneev, Yu. Korneeva, I. Florya, B. Voronov, and G. Goltsman 2012 *Physics Procedia* **36**, 72

[3] M. Sclafani, M. Marksteiner, F. Keir, A. Divochiy, A. Korneev, A. Semenov, G. Goltsman, and M. Arndt 2012 *Nanotechnology* **23**, 065501

[4] V. Kovalyuk, W. Hartmann, O. Kahl, N. Kaurova, A. Korneev, G. Goltsman, and W. H. P. Pernice 2013 *Opt. Express* **21**, 22683

[5] J. J. Renema, G. Frucci, Z. Zhou, F. Mattioli, A. Gaggero, R. Leoni, M. J. A. de Dood, A. Fiore, and M. P. van Exter 2012 *Opt. Express* **20**, 2806

[6] J. J. Renema, G. Frucci, M. J. A. de Dood, R. Gill, A. Fiore, and M. P. van Exter 2012 *Phys. Rev. A* **86**, 062113

[7] R. Hostein, R. Braive, M. Larqué, K.-H. Lee, A. Talneau, L. Le Gratiet,I. Robert-Philip, I. Sagnes, and A. Beveratos 2009 *Appl. Phys. Lett.* **94**, 123101

[8] T R. Hostein, A. Michon, G. Beaudoin, N. Gogneau, G. Patriache, J.-Y. Marzin, I. Robert-Philip, I. Sagnes and A. Beveratos 2008 *Appl. Phys. Lett.* **93**, 073106

[9] M. Halder, A. Beveratos, N. Gisin, V. Scarani, C. Simon, and H. Zbinden 2007 *Nat. Phys.* **3**, 692-695

[10] Marsili, F., Bellei, F., Najafi, F., Dane, A. E., Dauler, E. A., Molnar, R. J. & Berggren, K. K. 2012 *Nano Letters* **12**(9), 4799–4804

[11] G Gol'tsman, K Smirnov, P Kouminov, B Voronov, N Kaurova, V Drakinsky, J Zhang, A Verevkin and Roman Sobolewski 2003 *IEEE Transactions* **13**(2) pp192-195

[12] Li, H., Zhang, L., You, L., Yang, X., Zhang, W., Liu, X. and Xie, X. 2015 *Optics Express* **23**(13), 17301

[13] Kerman, A. J., Dauler, E. A., Keicher, W. E., Yang, J. K. W., Berggren, K. K., Gol'tsman, G. N & Voronov, B. 2006 *Appl. Phys. Lett.* **88**(11), 111116

[14] Baek, B., Lita, A. E., Verma, V., & Nam, S. W. 2011 *Appl. Phys. Lett.* **98**(25), 251105

[15] P. Zolotov, A. Divochiy, Y. Vakhtomin, V. Seleznev, P. Morozov, K. Smirnov Superconducting Single-photon Detectors Made of Ultra-thin VN Films, PhIO-2018, VII International Conference on Photonics and Information Optics, Volume 2018

[16] Gao, J. R., Hajenius, M., Tichelaar, F. D., Klapwijk, T. M., Voronov, B., Grishin, E. and Mehregany, M. 2007 *Appl. Phys. Lett.* **91**(6), 062504

[17] K. Smirnov, A. Divochiy, Y. Vakhtomin, P. Morozov, P. Zolotov, A. Antipov and V. Seleznev 2018 *Supercond. Sci. Technol.* **31**(3), 035011

[18] Seleznev V, Divochiy A, Vakhtomin Y, Morozov P, Zolotov P, Vasil'ev D, Moiseev K, Malevannaya E and Smirnov K 2016 *J. Phys.: Conf. Ser.* **737**, 012032

[19] Vasilev, D., Malevannaya, E. and Moiseev, K. 2017 *J. Phys.: Conf. Ser.* **872**, 012027

[20] Smirnov, K., Vachtomin, Y., Divochiy, A., Antipov, A. & Goltsman, G. 2015 *Appl. Phys. Express* **8**(2), 022501

Providing of Ultra-Thin Film Thickness Uniformity by Magnetron Sputtering from Two Sources

S Hydyrova[1], M Yu Akishin[1], D D Vasilev[1] and K M Moiseev[1]
[1] Bauman Moscow State Technical University, 105005, Moscow, Russia

E-mail: hydyrovas@yandex.ru, d.d.vasiliev@bmstu.ru, k.moiseev@bmstu.ru

Abstract. To achieve high characteristics of single-photon detectors based on WSi thin films it is necessary to provide high film thickness uniformity more than 95%. This paper presents a mathematical modeling of the thickness uniformity of an ultra-thin film formed by magnetron sputtering from two sources, based on an experimentally determined mass flow from a single magnetron, depending on the location of the magnetrons relative to the substrate holder. According to the results of non-uniformity modeling, the requirements for the design and location of the magnetron unit in the chamber were made, which ensured the required non-uniformity of the film thickness of less than 5%.

1. Introduction

Superconducting single photon nanowire detectors (SNSPD) are designed to detect visible radiation (single photons) (400 ... 780 nm) and near infrared (0.74 ... 2.5 microns). SNSPDs are used as photon counters in various fields: in quantum cryptography, in optical coherent tomography and quantum optical coherent tomography, in LIDAR systems, in astronomical research, and also in testing microcircuits [1]. The application of SNSPD in optical quantum computing is promising, since the implementation of a quantum optical integrated circuit requires a photon source, logic elements, and a photon receiver (detector) [2]. In addition, SNSPDs are in demand in optical systems for long-distance space communications [3].

Single-photon detectors based on superconductors have advantages over alternative ones due to a wider spectral range of detection, high quantum efficiency, high count rate and lower dark count rate, as well as higher operating temperatures [1, 4].

One of the most promising materials for the SNSPD sensitive element is WSi [5], the detectors based on which demonstrate high quantum efficiency up to 93% [6].

The thickness of the superconducting films for SNSPD is less than 10 nm [7]. At these values, a strong change is observed in the superconducting parameters of the material depending on the thickness [8], which leads to a significant effect of the film thickness on the output characteristics of the detector: quantum efficiency, count rate, operating temperature, and the number of dark counts [9, 10, 11]. In this regard, to ensure reproducibility of the output characteristics of the detector, it is necessary to obtain films with a high uniformity of thickness of more than 95%.

One of the most common methods for the formation of superconducting ultrathin films for SNSPDs, which are also used in the laboratory of the Department of Electronic Technologies in Mechanical Engineering, is magnetron sputtering. The formation of the WSi film in our laboratory is carried out at the VUP-11M technological system [12] by sputtering the W and Si targets of two magnetrons. The angle between the axis of each of the magnetrons and the substrate holder is 45°. The center of the substrate holder coincides with the intersection of the magnetron axes [12]. In [13] it was shown that the implementation of the rotation of the substrate holder at a given arrangement of magnetrons will provide the required non-uniformity of the film thickness of less than 5%. However, the current location of the substrate holder in the chamber (in a plane located at an angle to the axis of the vacuum input of the substrate holder) does not allow rotation of the substrate holder, and therefore the formed two-component films have a high non-uniformity of thickness and composition along the substrate.

To ensure film thickness non-uniformity of less than 5%, it is advisable to change the layout and place two magnetrons on one chamber flange symmetrically with respect to the substrate, and the substrate holder on the opposite flange, which will allow rotation of the substrate holder and reduce the non-uniformity of the film thickness and composition.

In this paper, we consider a method for ensuring the required non-uniformity of the thickness of a two-component film of less than 5% by means of the newly developed design of the magnetron assembly. To put forward the requirements for the design of the magnetron assembly (to determine the distance between the magnetron axes and the angle of inclination of the magnetron axis relative to the axis of the substrate holder), a non-uniformity of the film thickness by co-sputtering from two sources is simulated.

2. Experimental determination of the spatial distribution of the W flux from one magnetron

Modeling of thickness non-uniformity depending on the relative position of the magnetrons and the substrate holder is carried out on the basis of a regression model of the spatial distribution of the flux from one magnetron, obtained on the basis of an experimental measurement of the deposition rate.

To determine the material flux from a single magnetron, the mass deposited on quartz resonators fixed at 9 experimental points (Figure 1, a) on a substrate holder is determined using the piezoelectric microweighting method [14]. Tungsten W was deposited at five different distances z between the substrate holder and the magnetron: 50, 70, 90, 110, and 130 mm (Figure 1, b). The formation of a tungsten film by magnetron sputtering is carried out at the VUP-11M setup. The parameters of the deposition mode of the tungsten film [15] are shown in table 1.

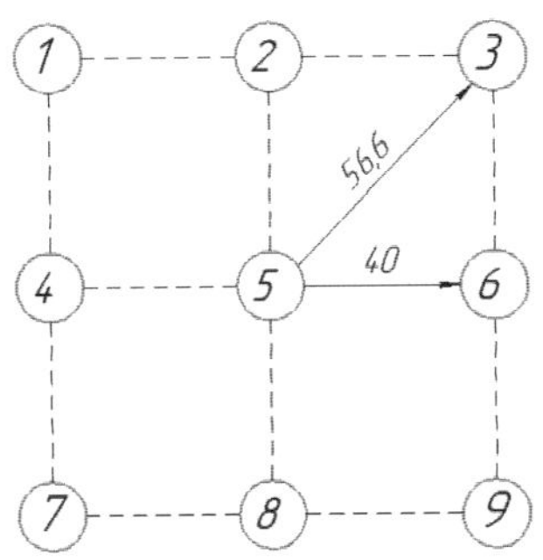

(a)

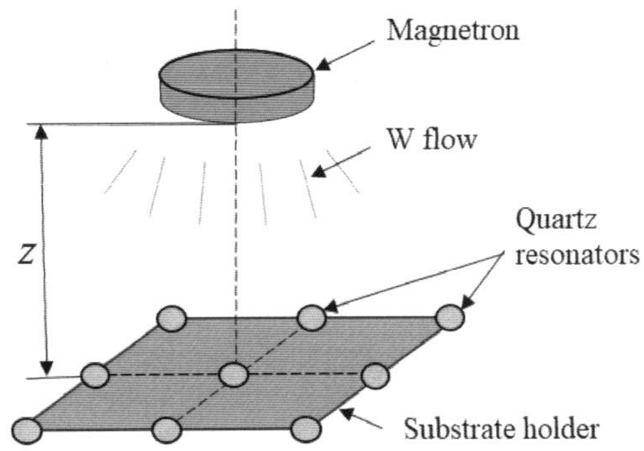

(b)

Figure 1. The arrangement of the experimental points: (a) the location of the substrate holder relative to the magnetron; (b) the location of the quartz resonators on the substrate holder.

Table 1. The parameters of the W films magnetron sputtering.

Parameter	Value
Signal	Pulse
Power (W)	120

IOP Conf. Series: Materials Science and Engineering **781** (2020) 012012 doi:10.1088/1757-899X/781/1/012012

Frequency (kHz)	15
Fill factor (%)	80
Argon flow (l/h)	1,56
Residual pressure (10^{-5} mbar)	1,1
Target training time (min)	5
Deposition time (min)	2

As a result of the experiments, the specific mass of the film deposited at each of the experimental points was calculated. The flux Q [g/(cm^2·s)] is determined from the ratio of the specific gravity to the film deposition time t, the calculated flux values at the experimental points for the distances z from the magnetron are presented in Table 2.

Table 2. Values of the tungsten flow Q [g/(cm2·s)] at the experimental points.

Resonator No	z_1 = 50 мм	z_2 = 70 мм	z_3 = 90 мм	z_4 = 110 мм	z_5 = 130 мм
1	0,491	0,339	0,337	0,274	0,187
2	0,984	0,447	0,478	0,309	0,245
3	0,536	0,354	0,444	0,252	0,240
4	0,884	0,618	0,492	0,415	0,279
5	2,086	0,958	0,763	0,450	0,339
6	0,808	0,607	0,534	0,286	0,289
7	0,537	0,501	0,417	0,321	0,248
8	0,799	0,603	0,570	0,322	0,285
9	0,510	0,395	0,386	0,219	0,236

A mathematical model of the dependence of the material flow on spatial coordinates relative to the magnetron was obtained on the basis of the experimental values of the flux Q using regression analysis in a Mathcad environment.

Using the third-order polynomial regression function, we obtained the regression models of the flow Q depending on the location of the point on the substrate holder in the Cartesian coordinates x, y. The coordinates of the extrema of the functions were found to eliminate the influence of the error of fixing the central resonator on the substrate holder (mismatch between the center of the resonator 5 and the axis of the magnetron (figure 1 a, b). Then the coordinates of the experimental points 1 ... 9 for each fixing height of the substrate holder z are translated into polar coordinates r and φ, and flux values Q are averaged over the angle φ according to the mean value theorem. Thus, we obtain the functions of the flux $Q(r)$ depending on the radius relative to the magnetron axis for each height.

The obtained $Q(r)$ curves at heights z_1 - z_5 are combined by the regression function into a model of the spatial distribution of the material flow Q at a given point from one magnetron depending on the radius r and height z (figure 2), where r and z are the radius and height in cylindrical coordinates relative to the magnetron. The location of the magnetron corresponds to the origin on the graphs.

The obtained spatial distribution of the material flow from one magnetron Q allows us to simulate the dependence of the non-uniformity of the film thickness on the relative position of the substrate and magnetrons during sputtering from two sources.

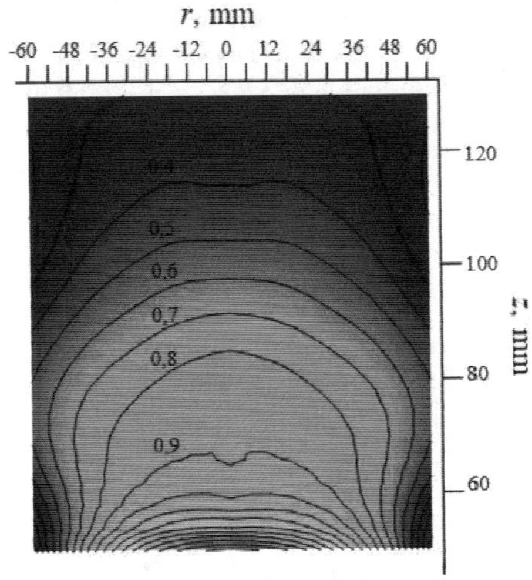

(a)

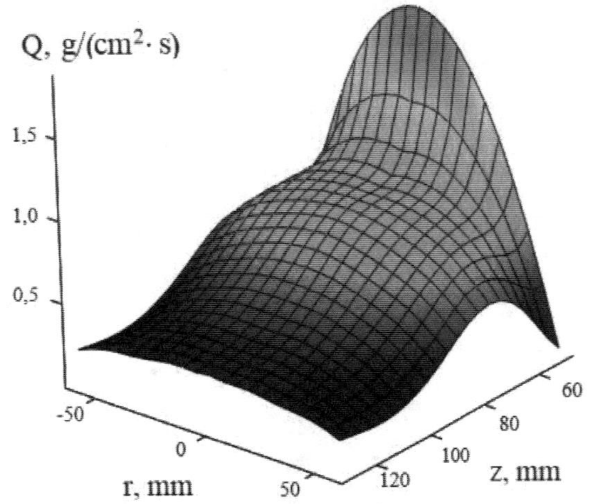

(b)

Figure 2. Regression model of the dependence of the flux *Q* [g/(cm^2·s)] on the radius *r* and height *z* (the location of the magnetron corresponds to the origin): a) contour plot; b) surface graph.

3. Modeling film thickness non-uniformity depending on the magnetron location relative to the substrate

To simulate the film thickness non-uniformity depending on the substrate location relative to the magnetron, the coordinates *r* and *z* are expressed in terms of angle β and distances *B* and *L* (figure 3).

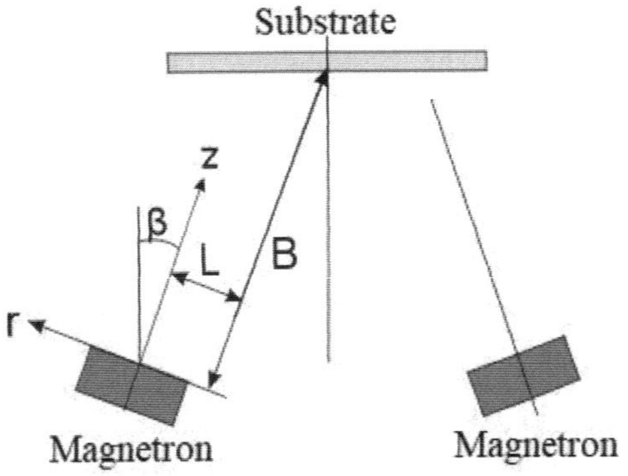

Figure 3. The calculational scheme of the substrate layout relative to the magnetrons.

Based on the flow distribution, a film thickness non-uniformity model Δ and a specific film deposition rate (which is equivalent to *Q* flow) at sputtering from two sources were built in Mathcad depending on *L* and *B* at an angle β from 0 to 45° in increments of 5° with ranges of variation of *L* from 0 to 40 mm, and of *B* from 90 to 130 mm. From 10 obtained plots of the non-uniformity surface Δ depending on the distances L and B, it was found that the smallest angle β, at which Δ is less than 5%, is equal to 5°. With an increase in β, the ranges *B* and *L*, in which the condition $\Delta < 5\%$ is fulfilled, increase, and these regions shift towards lower values of *B*. Based on the plots of the deposition rate versus *B* and *L* for various β, it was found that the effect of angle β on the deposition rate is negligible. The deposition rate decreases with increasing distance *B* and eccentricity *L*.

In addition to the film thickness non-uniformity $\Delta < 5\%$, the arrangement of magnetrons in the chamber should provide a high film deposition rate. The angle of magnetron inclination β should be the smallest in order to place the magnetron assembly in the chamber of the VUP-11M system. Based on these requirements and the simulation of non-uniformity and deposition rate, the angle $\beta = 10°$ was chosen. At $\beta = 10°$, non-uniformity of less than 5% is ensured with a distance B = 105...120 mm and an eccentricity L = 0...5 mm. In the selected range, the highest deposition rate of 0.496 g/(cm2·s) is achieved at L = 0 mm and B = 105 mm. Models of non-uniformity of the film thickness Δ [%] and deposition rate Q [g/(cm2·s)] at β = 10 ° depending on L and B are presented in figures 4 and 5, respectively.

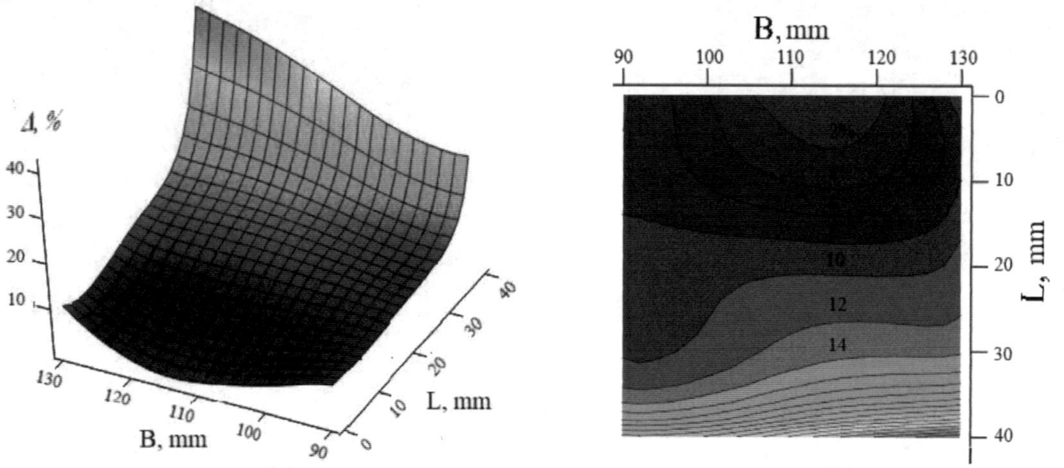

(a) (b)

Figure 4. Model of the film thickness non-uniformity Δ [%] depending on the distance B from the substrate holder and eccentricity L at magnetron inclination angle β = 10°: a) surface graph; b) contour plot.

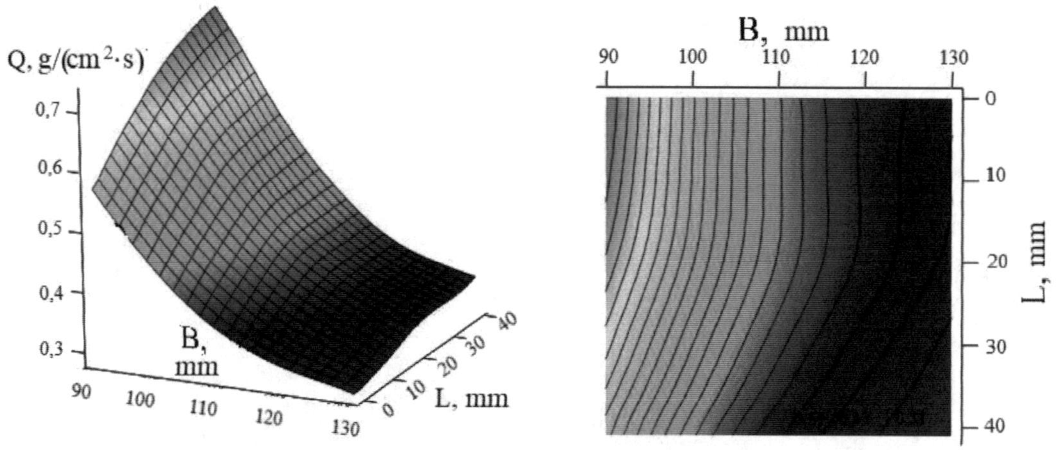

(a) (b)

Figure 5. Model of the film deposition rate Q [g/(cm^2·s)] depending on the distance B from the substrate holder and eccentricity L at magnetron inclination angle β = 10°: a) surface graph; b) contour plot.

Based on the modeling of the thickness and deposition rate of the film at Mathcad, the following requirements for the location of the magnetrons relative to the substrate holder, which provide the film thickness non-uniformity is less than 5%, have been put forward:

- magnetron inclination angle $\beta > 10°$
- eccentricity $L < 5$ mm
- distance B =105…130 mm

According to the results of modeling the film deposition rate for $\beta = 10°$ in the indicated ranges of L and B, the maximum deposition of 0.496 g/(cm^2·s) is provided at $L = 0$ mm and $B = 105$ mm. Based on the above requirements, k and h are chosen (figure 6):

$$h = B\cos\beta \ L\sin\beta = 105 \cdot \cos(10°) = 103{,}4 \approx 104\,mm \quad (1)$$

$$k = B\sin\beta + L\cos\beta = 105 \cdot \sin(10°) = 18{,}2 \approx 20\,mm \quad (2)$$

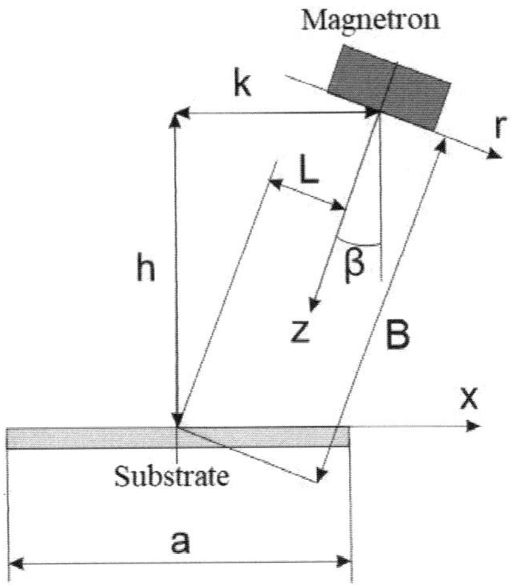

Figure 6. The calculational scheme of the magnetron location relative to the substrate.

Thus, based on the requirements for thickness non-uniformity of less than 5%, a high film deposition rate and a small angle of inclination, the distances from the substrate holder to the magnetron are chosen to be k = 20 mm and h = 104 mm at an angle β = 10 °, which satisfies the required conditions for ensuring thickness non-uniformity films less than 5%. This arrangement of the magnetron assembly can be implemented at the VUP-11M installation for the formation of two-component WSi thin films for single-photon detectors.

4. Conclusions

Based on the simulation of the WSi film thickness non-uniformity depending on the location of the magnetrons relative to the substrate holder, it was determined that the required non-uniformity of the thickness of less than 5% of the film is ensured at $\beta = 10°$, distance B = 105...120 mm and eccentricity L = 0...5 mm. In the selected range, the highest deposition rate of 0.496 g/(cm^2·s) is achieved at $L = 0$ mm and B = 105 mm, which corresponds to the distance between magnetrons $2k$ = 40 mm, the distance between magnetrons and substrate holder h = 104 mm at an angle between the magnetron axes and substrate holder $\beta = 10°$. This arrangement of the magnetron assembly can subsequently be implemented on the VUP-11M system for the formation of two-component WSi thin films for single-photon detectors.

References

[1] Minaeva O V et al 2013 High-speed single-photon detector based on a superconducting NbN thin film (Moscow: Prometey) (in Russian)

[2] Smirnov K V 2013 Features of heating and relaxation of hot electrons in thin-film superconducting nanostructures and 2D semiconductor heterostructures upon absorption of radiation from the infrared and terahertz ranges *Dissertation (PhD)* Moscow Pedagogical State University (in Russian)

[3] Sanginetti B, Amri E and Richow F 2017 *Photonics* 3 56-69

[4] Kitaygorsky J 2008 Photon and dark counts in NbN superconducting single-photon detectors and nanostripes *Doctoral dissertation* New York

[5] Zhang X 2018 Characteristics of tungsten silicide and its application for single X-ray photon detection *Doctoral dissertation* University of Zurich

[6] Marsili F, Verma V B, Stern J A, Harrington S, Lita A E, Gerrits T, ... & Nam S W 2013 *Nature Photonics*, **7**(3) 210

[7] Jin J, Fu F, Jia X, Kang L, Wang Z, Tu X, ... & Wu P 2019 *IEEE Transactions on Applied Superconductivity* **29**(5) 18483128 1-4

[8] Zhang X, Engel A, Wang Q, Schilling A, Semenov A, Sidorova M, ... & Siegel M 2016 *Physical Review B* **94**(17) 174509

[9] Murphy R, Grein M, Gudmundsen T, McCaughan A, Najafi F, Berggren K K, ... & Dauler E 2015 In *CLEO: QELS_Fundamental Science* (pp. FF2A-3) Optical Society of America

[10] Ren X 2015 Advanced photon counting techniques for long-range depth imaging *Doctoral dissertation* Heriot-Watt University

[11] Banerjee A, Baker L J, Doye A, Nord M, Heath R M, Erotokritou K, ... & Hadfield R H 2017 *Superconductor Science and Technology* **30**(8) 084010

[12] Vasilev D, Malevannaya E and Moiseev K 2017 *J. of Phys.: Conf. Ser.* **872**(1) (IOP Publishing) 012027

[13] Vasiliev D D, Malevannaya E I, Moiseev K M 2015 *Materials of the 22th Scientific and Technical Conf. with the participation of foreign specialists "Vacuum Science and Technology"* ed A S Bugaev (Moscow: NOVELLA) pp18-22 (in Russian)

[14] Sauerbrey G 1959 *Zeitschrift für physik* **155**(2) pp206-22 (in German)

[15] Seleznev V, Divochiy A, Vakhtomin Yu, Morozov P, Zolotov P, Vasil'ev D, Moiseev K, Malevannaya E and Smirnov K 2016 *J. of Phys.: Conf. Ser.* **737** 012032

IOP Conf. Series: Materials Science and Engineering **781** (2020) 012013 doi:10.1088/1757-899X/781/1/012013

Influence of Deposited Material Energy on Superconducting Properties of the WSi Films

D D Vasilev[1], E I Malevannaya[1], K M Moiseev[1], P I Zolotov[2,3], A V Antipov[2,3], Y B Vakhtomin[2,3] and K V Smirnov[2,3,4]

[1] Bauman Moscow State Technical University, 105005, Moscow, Russia
[2] Moscow Pedagogical State University, 119991, Moscow, Russia
[3] LLC Superconducting nanotechnology (Scontel), 119021, Moscow, Russia
[4] National Research University Higher School of Economics, 101000, Moscow, Russia

E-mail: d.d.vasiliev@bmstu.ru, k.moiseev@bmstu.ru

Abstract. WSi thin films have the advantages for creating SNSPDs with a large active area or array of detectors on a single substrate due to the amorphous structure. The superconducting properties of ultrathin WSi films substantially depends on their structure and thickness as the NbN films. Scientific groups investigating WSi films mainly focused only on changes of their thickness and the ratio of the components on the substrate at room temperature. This paper presents experiments to determine the effect of the bias potential on the substrate, the temperature of the substrate, and the peak power of pulsed magnetron sputtering, which is the equivalent of ionization, a tungsten target, on the surface resistance and superconducting properties of the WSi ultrathin films. The negative effect of the substrate temperature and the positive effect of the bias potential and the ionization coefficient (peak current) allow one to choose the best WSi films formation mode for SNSPD: substrate temperature 297 K, bias potential −60 V, and peak current 3.5 A.

1. Introduction

Superconducting single photon nanowire detector (SNSPD) was developed in 2001 [1]. Such detectors are used as a single photon counter for secure quantum keys' distribution in quantum cryptography [2], in optical space communications [3], for CMOS microcircuit testing [4], visualizing the depth of a subject with high resolution over long distances [5], creating a short-wave infrared single-photon camera [6] etc. To introduce radiation into the meander, a single-mode optical fiber is used, the core thickness of which varies from 9 to 11 μm, so the active region of the meander was initially from 10 to 11 μm, while now it has increased to 14-16 μm to reduce radiation losses [7-11]. If there is an airspace between the photon source and the detector, then it is more efficient to use multimode optical fiber, for example, for use in singlet oxygen luminescence dosimetry (SOLD) [12]. The core diameter of a multimode optical fiber starts at 50 μm, which requires the use of detectors with an active region of about 35 μm in diameter and focusing lenses [13], or detectors with an active region of about 50 μm [14]. Using a matrix of detectors will increase the image parameters of cameras based on single-photon detectors [15]. The demonstrated possibilities of integrating the detector into photonic [16] and plasmonic [17] schemes require the creation of a detectors' array on one substrate with high output parameters [18].

Initially, the detector sensitive element, the meander, was formed on the basis of ultrathin polycrystalline NbN films [19]. Due to the structure and properties of such films are extremely sensitive to the substrate material and the formation conditions, as a result up to now it has not been possible to reproducibly form NbN films on relatively large areas of the substrate. Detectors based on amorphous films, such as WSi [20], MoSi [21], and MoGe [22], are more promising due to their significantly less dependent on the substrate material and its surface structure, as well as less critical to defects in the film itself [23]. Detectors based on amorphous WSi films show high quantum efficiency up to 93% at a telecommunication wavelength of 1550 nm [24]. In addition, due to the smaller band gap and structural homogeneity, detectors based on WSi films with an amorphous structure can effectively register mid-infrared photons with a standard nanowire width of about 100 nm [20], while for detecting mid-infrared photons by means of detectors based on NbN films it is required to decrease the width of the nanowire up to 30 nm, which complicates the manufacturing process [25].

The parameters of the ready-made detector depend on the properties of the ultrathin film of the sensitive element, which in turn depend on its structure. WSi films transfer to the superconducting state at a temperature of about 5 K, the amorphous structure and silicon content of 25% [26]. In this case, the scientific groups during the development of the technology for the formation of WSi films vary the ratio of the components and the film thickness, depositing onto the substrate at room temperature [20, 27-30]. It was noted that WSi films have a lower critical current on MgO substrates in comparison with SiO_x and Si, while the temperature of the transition to the superconducting state on these substrates coincides [30]. An analysis of X-ray diffraction (XRD) showed a non-intense and wide peak of the WSi_2 (110) and W_5Si_3 (202) gratings. Measurements of the film resistance upon application of a magnetic field indicate the anisotropic properties of thin WSi films that are not characteristic of the amorphous structure, which may indicate the presence of a preferred orientation in WSi films [30]. For NbN films, studies were carried out on the effect of the bias potential on the substrate [31], the temperature of the substrate in the region of 825°C [32] and room temperature [33], ion stimulation [34] on the structure and superconducting properties, while the data of the deposited material energy effect on the structure and properties of WSi films are absent. In this paper, we present the results of a study of the magnetron sputtering process technological regimes influence on the superconducting properties of the WSi films.

2. Experiment description

According to the zone model of the film structure for an unbalanced magnetron system, the film structure from zone 1, which is characterized by cone-shaped crystallites separated by voids, transfers to zone 2 (close-packed columnar grains) and 3 (recrystallized grains) with an increase in the substrate temperature, the bias potential on the substrate, and the flux of ions incident on the substrate with respect to the flux of condensing atoms on the substrate (Figure 1) [35]. A study of the influence of all three factors was implemented at the VUP-11M vacuum system at BMSTU, upgraded for WSi films forming process from two magnetron sputtering sources [36]. In our experiment, the substrate temperatures *Ts* are 297, 447, and 597 K, and the bias potential *Ubias* is 0, −30, −60 V. During the formation of WSi films, silicon is sputtered in the high-frequency mode; therefore, it is possible to control only the ionization of W sputtered in the pulsed mode. The degree of W ionization correlates with the peak current in the pulse *Ipeak*, the values of which are 0.5, 2.0, and 3.5 A, based on previous studies [37].

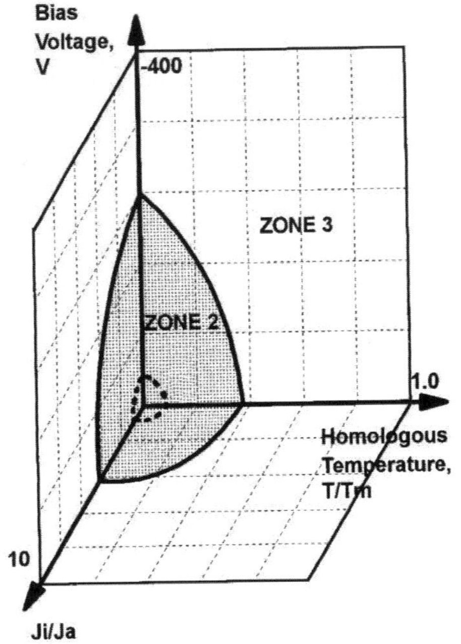

Figure 1. A zone model of the film structure for an unbalanced magnetron system [35].

Influence evaluation of each factor separately and their interaction was carried out according to the full factorial experiment method (Figure 2). To estimate the influence of each factor individually, three additional experiments are carried out at points with a middle value of the varied factors. The experiments are carried out at a working pressure of $2.1 \cdot 10^{-3}$ mbar, an Ar flow of 2.6 l/h, a deposition time of 15 s, a concentration of W of 76%, a specific weight of the film of $8.966 \cdot 10^{-5}$ kg/m^2, which, according to estimates, corresponds to a thickness of 6 nm.

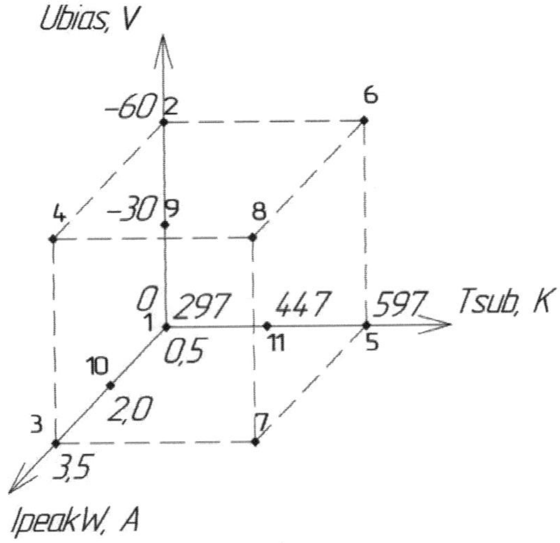

Figure 2. Full factorial experiment scheme.

It should be noted that at film thicknesses less than 10 nm, the thickness value significantly affects their surface resistance and superconducting properties [30]. Due to the influence of the studied factors on the structure, and, consequently, on the density of the formed films, which will change the thickness of the films at the same deposition rate, it was proposed to form films with the same specific weight, which is controlled by the piezoelectric microweighting method according to the previously developed technique [38].

The surface resistance *Rs* was measured for the obtained WSi films by the four-probe method, and the superconducting properties when the sample was lowered into the pumped-out insert in the cryostat were as follows: critical temperature *Tc*, transition width *ΔTc*, and residual-resistance ratio *RRR*.

3. Results and discussion

The surface resistance *Rs* of the WSi films increases with increasing substrate temperature *Ts* and the bias potential *Ubias*. Peak current *Ipeak* does not affect surface resistance (Figure 3).

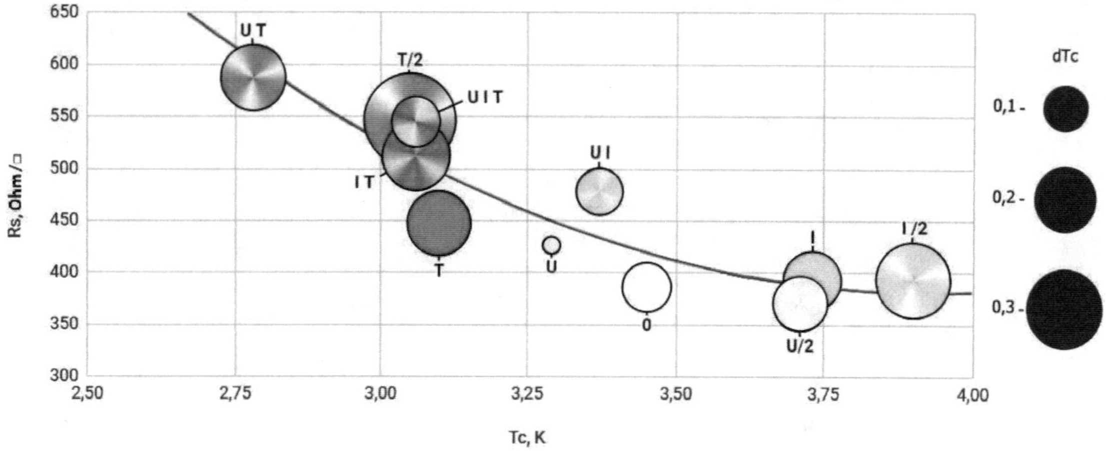

Figure 3. The dependence of surface resistance on the transition temperature of the film, the size is the width of the transition.

The transition temperature *Tc* decreases with increasing substrate temperature *Ts* and the bias potential *Ubias*, and increases with increasing peak current *Ipeak*.

Variable factors do not affect the residual-resistance ratio *RRR* (table 1). The dependence of the width of the transition to the superconducting state *dTc* on varied factors was also not detected (Figure 3).

Table 1. Surface resistance and superconducting properties of samples.

Sample No	Factors	*Rs* (Ohm/□)	*Tc* (K)	*dTc* (K)	*RRR*
1	–	387	3.45	0.12	0.95
2	U	426	3.29	0.07	0.96
3	I	392	3.73	0.17	0.94
4	U I	478	3.37	0.12	0.96
5	T	447	3.10	0.20	0.96
6	U T	587	2.78	0.20	0.96
7	I T	513	3.06	0.24	0.95
8	U I T	545	3.06	0.13	0.96
9	U/2	370	3.71	0.15	0.96
10	I /2	393	3.90	0.26	0.95
11	T/2	546	3.05	0.38	0.95

As a result, an increase in the temperature of the substrate increases the surface resistance of the film and decreases the critical temperature, which may be associated with the formation of undesirable bonds between particles of the deposited material and elements of the residual atmosphere or substrate material.

An increase in the bias potential increases the surface resistance without changing the superconducting properties of the film, which can improve the detector signal-to-noise ratio.

An increase in the peak current during W sputtering increases the critical temperature without changing the surface resistance, and can lead to an increase in the critical current density, which can have a positive effect on the detector parameters: signal-to-noise ratio and pulse generation efficiency.

The residual-resistance ratio *RRR* of the WSi films, unlike NbN, varies from 0.94 to 0.96, which does not allow us to consider this parameter as a tool to improve the output characteristics of the detector.

4. Conclusions

The formation of the WSi films at a substrate temperature of 297 K, a bias potential of −60 V, and a peak current of W 3.5 A complexly increases the surface resistance of the films Rs and the critical transition temperature to the superconducting state Tc, which can positively affect the improvement of the detector signal-to-noise ratio, and contribute to the improvement of the impulse generation of the films. However, to fully estimate the influence of factors on the detector parameters, it is necessary to study the influence of factors on the critical current density and the detector output parameters.

5. Acknowledgment

The research is supported by the Russian Science Foundation project No. 18-12-00364.

References

[1] Gol'tsman G, Okunev O, Chulkova G, Lipatov A, Semenov A, Smirnov K, Voronov B, Dzardanov A, Williams C and Sobolewski R 2001 *Appl. Phys. Lett.* **79**, 705

[2] Hoi-Kwong L, Curty M and Tamaki K 2014 *Nat. Photonics* **8**, 595–604

[3] M. Grein, E. Dauler, A. Kerman, M. Willis, B. Romkey, B. Robinson, D. Murphy, and D. Boroson, "A superconducting photoncounting receiver for optical communication from the Moon," SPIE, 2015

[4] Zhang J, Boiadjieva N, Chulkova G, Deslandes H, Gol'tsman G, Korneev A, Kouminov P 2003 *Electronics Letters* **39**(14) 1086

[5] Aongus McCarthy, Nils J. Krichel, Nathan R. Gemmell, Ximing Ren, Michael G. Tanner, Sander N. Dorenbos, Val Zwiller, Robert H. Hadfield and Gerald S. Buller 2013 *Opt. Express* **21**, 8904

[6] Thomas Gerrits, Daniel J. Lum, Varun Verma, John Howell, Richard P. Mirin, and Sae Woo Nam 2018 *Opt. Express* **26**, 15519

[7] Iman Esmaeil Zadeh, Johannes W. N. Los, Ronan B. M. Gourgues, Violette Steinmetz, Gabriele Bulgarini, Sergiy M. Dobrovolskiy, Val Zwiller, and Sander N. Dorenbos 2017 *APL Photonics* **2**, 111301

[8] Lixing You, Jia Quan, Yong Wang, Yuexue Ma, Xiaoyan Yang, Yanjie Liu, Hao Li, Jianguo Li, Juan Wang, Jingtao Liang, Zhen Wang, and Xiaoming Xie 2018 *Opt. Express* **26**, 3 2965

[9] Li Chen, Dirk Schwarzer, Jascha A. Lau, Varun B. Verma, Martin J. Stevens, Francesco Marsili, Richard P. Mirin, Sae Woo Nam, and Alec M. Wodtke 2018 *Opt. Express* **26**, 12 14859

[10] Francesco Bellei, Alyssa P. Cartwright, Adam N. McCaughan, Andrew E. Dane, Faraz Najafi, Qingyuan Zhao, and Karl K. Berggren 2016 *Opt. Express* **24**, 4 3248

[11] P. Zolotov, A. Divochiy, Yu. Vakhtomin, M. Moshkova, P. Morozov, V. Seleznev, and K. Smirnov 2018 *AIP Conference Proceedings* **1936**, 020019

[12] Nathan R. Gemmell, Aongus McCarthy, Michele M. Kim, Israel Veilleux, Timothy C. Zhu, Gerald S. Buller, Brian C. Wilson, and Robert H. Hadfield 2017 *J. Biophotonics* **10,** 2, 320

[13] Dengkuan Liu, Shigehito Miki, Taro Yamashita, Lixing You, Zhen Wang, and Hirotaka Terai 2014 *Opt. Express* **22**, 18, 21167

[14] Hao Li, Lu Zhang, Lixing You, Xiaoyan Yang, Weijun Zhang, Xiaoyu Liu, Sijing Chen, Zhen Wang, and Xiaoming Xie 2015 *Opt. Express* **23**, 13, 17301

[15] T. Gerrits, S. Allman, D. J. Lum, V. Verma, J. Howell, R. Mirin, and S. W. Nam, "Progress toward a high-resolution single-photon camera based on superconducting single photon detector arrays and compressive sensing," in CLEO: 2015, OSA Technical Digest (online) (Optical Society of America, 2015), paper STh3O.6
[16] F. Lenzini, N. Gruhler, N. Walter, W. H. P. Pernice 2018 *Adv. Quantum Technol.* **1**, 1800061
[17] Reinier W. Heeres, Leo P. Kouwenhoven & Valery Zwiller 2013 *Nat. Nanotechnology* **8**, 719
[18] M. S. Allman, V. B. Verma, M. Stevens, T. Gerrits, R. D. Horansky, A. E Lita, F. Marsili, A. D. Beyer, M. D. Shaw, D. Kumor, R. Mirin, and S. W. Nam 2015 Proc. SPIE 9504, *Photon Counting Applications* 2015, 950402
[19] J. R. Gao and M. Hajenius 2007 *Appl. Phys. Lett.* **91**, 062504
[20] Burm Baek, Adriana E. Lita, Varun Verma, and Sae Woo Nam 2011 *Appl. Phys. Lett.* **98**, 251105
[21] Korneeva Y P, Mikhailov M Y, Pershin Y P, Manova N N, Divochiy A V, Vakhtomin Y B, and Goltsman G N 2014 *Superconductor Science and Technology* **27**, 095012
[22] V. B. Verma, A. E. Lita, M. R. Vissers, F. Marsili, D. P. Pappas, R. P. Mirin, and S. W. Nam 2014 *Appl. Phys. Lett.* **105**, 022602
[23] Lita, A., Verma, V., Horansky, R., Shainline, J., Mirin, R., & Nam, S. (2015). Materials Development for High Efficiency Superconducting Nanowire Single-Photon Detectors. MRS Proceedings, 1807, 1-6
[24] Marsili F, Verma V B, Stern J A, Harrington S, Lita A E, Gerrits T and Nam S W 2013 *Nature Photonics* **7**(3) 210
[25] Francesco Marsili, Francesco Bellei, Faraz Najafi, Andrew E. Dane, Eric A. Dauler, Richard J. Molnar, Karl K. Berggren 2012 *Nano Lett.* 20121294799-4804
[26] Kondo S 1992 *Journal of materials research* **7**(04), 853
[27] X. Zhang, A. Engel, Q. Wang, A. Schilling, A. Semenov, M. Sidorova, H.-W. Hübers, I. Charaev, K. Ilin, and M. Siegel 2016 *Phys. Rev.* B **94**, 174509
[28] V A Seleznev, A V Divochiy, Yu B Vakhtomin, P V Morozov, P I Zolotov, D D Vasil'ev, K M Moiseev, E I Malevannaya and K V Smirnov 2016 *J. Phys.: Conf. Ser.* **737** 012032
[29] I. N. Florya, Yu. P. Korneeva, M. Yu. Mikhailov, A. Yu. Devizenko, A. A. Korneev, and G. N. Goltsman 2018 *Low Temperature Physics* **44**, 221
[30] Jin Jin, Fengfeng Fu, Xiaoqing Jia, Lin Kang, Zhihe Wang, Xuecou Tu, Labao Zhang 2019 *IEEE Transactions on Applied Superconductivity* **29**, 5, 18483128
[31] Andrew E. Dane, Adam N. McCaughan, Di Zhu, Qingyuan Zhao, Chung-Soo Kim, Niccolo Calandri, Akshay, Agarwal, Francesco Bellei, and Karl K. Berggren 2017 *Appl. Phys. Lett.* **111**, 12260
[32] K Ilin et al 2008 *J. Phys.: Conf. Ser.* 97 012045
[33] Zhen Wang, Akira Kawakami, Yoshinori Uzawa, and Bokuji Komiyama 1996 *J Appl. Phys.* **79**, 7837
[34] Tomas Polakovic, Sergi Lendinez, John E. Pearson, Axel Hoffmann, Volodymyr Yefremenko, Clarence L. Chang, Whitney Armstrong, Kawtar Hafidi, Goran Karapetrov, and Valentine Novosad 2018 *APL Materials* **6**, 076107
[35] P. J. Kelly and R. D. Arnell 1998 *Journal of Vacuum Science & Technology* A **16**, 2858
[36] D D Vasilev et al 2017 *J. Phys.: Conf. Ser.* **872** 012027
[37] Akishin M Y, Malevannaya E I, Vasiliev D D and Moiseev K M 2017 *Materials of the 24th Scientific and Technical Conf. with the participation of foreign specialists "Vacuum science and technology"* ed S B Nesterov (Moscow: NOVELLA) 193-6 (in Russian)
[38] Vasiliev D D, Malevannaya E I and Moiseev K M 2018 *25th Scientific and Technical Conf. with the participation of foreign experts "Vacuum Science and Technology"* ed S B Nesterov (Moscow: NOVELLA) 221-7 (in Russian)

IOP Conf. Series: Materials Science and Engineering **781** (2020) 012014 doi:10.1088/1757-899X/781/1/012014

Possibilities of the Electron-Beam Machine «LUCH» for Dimensional Microprocessing of Glass and Ceramic Materials

Y Zhuo[1], M Liang[1], K M Moiseev[1], L L Kolesnik[1] and Yu V Panfilov[1]
[1] Bauman Moscow State Technical University, 105005, Moscow, Russia
E-mail: k.moiseev@bmstu.ru, kolesnik@bmstu.ru, panfilov@bmstu.ru

Abstract. The article presents the modern applications and advantages of electron beam technology in the processing of glass and ceramic materials. The design and technological capabilities of the vacuum machine «LUCH» for electron beam melting, welding and dimensional microprocessing of vacuum glass and ceramic materials, including LTCC ceramics, are considered.

1. Introduction

The technology of electron beam processing has been known and has been used for more than 60 years, but interest from science and industry continues to grow [1]. One of the most interesting, but still little used applications, is micro- and nanoscale processing [2-4].

In the electronic industry and instrumentation, vacuum and optical glass and ceramic materials are used in the manufacture of parts and components of electronic devices. But the technology of melting, welding and dimensional processing of electric vacuum glass does not always meet modern requirements. For example: when drilling ceramic materials, including low-temperature co-fired ceramics (LTCC) used for the manufacture of microwave device substrates, by traditional methods (mechanical and laser methods), the hole diameter is up to 50 microns [5], however, for further miniaturization of electronic devices sizes should decrease to 30 microns and less [6].

LTCC has been used in various industries for many years. This ceramic is used as a substrate for microwave devices used to create microwave emitting devices, for example, Bluetooth and Wi-Fi modules. Improvements in LTCC materials, processes and manufacturing methods have led to lower costs and improved technical characteristics of electronic products. This has significantly increased interest in LTCC technology from manufacturers of high-frequency technology, optoelectronics and micro-electromechanical systems. New opportunities are opening up for the production of electronic products in such areas as telecommunications, medicine, and automotive technology [7].

Compared to mechanical processing, electron beam processing is devoid of such disadvantages as changing the size of the holes when the tool is worn and the need to change the tool when changing the profile of the hole, and compared to the laser processing, the lack of tapering of the holes and changing the composition of the material in the area near the processing [8-11] .

The main difficulties encountered in the electron beam treatment of dielectrics are associated with the low electrical conductivity of these materials [12, 13]. A low dielectric conductivity leads to charge accumulation, which reduces the electron energy of the beam, causing defocusing and distortion of the shape, as well as an increase in the diameter of the beam. Generation of an electron beam in the pressure range from 5 to 20 Pa leads to the formation of a plasma in the processing zone, which ensures the discharge of charge from the treated dielectric object [14, 15].

In addition to dimensional micro- and nano-processing, high-energy electron beams can be successfully used for other types of processing of glass and ceramic materials. The use of electron-beam heating during cutting and welding of parts from electrovacuum glass and ceramic materials ensures a constant chemical composition of the material in the zone of the welding joint. When melting, electron-beam heating eliminates the influence of the environment on the composition and structure of the material, and makes it possible to more precisely control its properties. Electron beam polishing creates additional opportunities for surface modification with a significant reduction in the surface roughness of the part [16].

For the implementation of processing technologies for products from vacuum glass and ceramic materials, as well as for further research on micro- and nanoscale processing of LTTC ceramics, the basic requirements for the basic processing technology and installation for their implementation are presented (Tables 1 and 2).

Table.1 Basic technology requirements.

Parameter	Value
Types of glass processing	Welding, smelting, dimensional processing
Product form	Tubular, flat
Welding speed, cm / min.	1.0
Penetration depth, mm	Up to 5.0

Table.2 Machine requirements.

Parameter	Value
Overall dimensions of installation, m	2.0 × 2.0 × 3.0
Vacuum chamber volume, m^3	0.3
Power consumption, kW	10.0
Control	Automated

2. Electron-beam machine

The «LUCH» technological machine (Figure. 1) includes a vacuum chamber with pumping and control system mounted on a single frame; in-chamber equipment for securing, rotating and moving processed products; an electronic source (gun) and an electronic source control unit.

Figure 1. Electron beam installation «LUCH»

The vacuum system of the installation is equipped with oil-free pumping means and provides a maximum pressure in the chamber of $5.0.10^{-4}$ Pa. The vacuum system consists of the following elements: Edwards XDS35i pump; Edwards EH250 pump; Edwards STPXA4503C ISO320F pump; shutter VAT Series 14 DN320 (14050-PE44); four valves VAT Series 26 DN63 (26436-QE41); three valves VAT Series 26 DN25 (26428-KE41); wide-range pressure transmitter Edwards WRG-S; pressure sensor thermocouple Edwards APG-100; control cabinet. Technical characteristics of the installation are given in Table 3.

Table.3 Technical characteristics of the electron-beam machine.

Parameter	Value
The volume of the vacuum chamber, m^3	0.3
Maximum residual pressure, Pa:	
- when using fore-vacuum pumps;	5.0×10^{-1}
- when using a high vacuum pump	5.0×10^{-4}
The total flow of leakage and gas separation, Pa × m^3 / s	5.0×10^{-3}
Number of process gas supply channels	2
Process gas flow rate for each channel, sccm	10.0 – 200.0
Range of working pressures, Pa	1.0 – 25.0
Pumping time to a pressure of 5.0×10^{-1} Pa, min	5
Pumping time to a pressure of 5.0×10^{-4} Pa, min	15

The vacuum chamber has a water-cooling system for the walls, upper and lower plates. On the top plate of the working chamber there is a flange for installing an electron beam gun and a device for its vertical movement. The camera door has a water-cooling system. A window with a diameter of 200 mm is located on the door of the chamber, made of materials that provide protection against ionizing radiation.

The control module contains control elements: a vacuum pump start-up system, a control and regulation system for valves. The main part of the control module is a specialized industrial controller that provides Ethernet data transfer between the server and the control units of the executing mechanisms.

The electrical equipment of the installation consists of pumping control systems, measuring pressures at characteristic points of the installation, supplying and controlling an electron-beam source, admitting process gases into the volume of the vacuum chamber, positioning the workpieces / products inside the vacuum chamber relative to the electron-beam source, video surveillance of the technological process, and also centralized control and visualization of the parameters of the installation.

One of the advantages of plasma sources of electron beams is the possibility of their operation in medium vacuum, however, the developed technology does not imply a change in the chemical composition of the material in the processing zone. Based on this, it becomes necessary not only to pump the working volume to the required pressure, but also to ensure that the composition of the residual atmosphere in the working chamber is replaced with an inert gas. For this purpose, a gas supply system with flow controllers is provided in the product design.

The gas inlet system in the volume of the vacuum chamber consists of two identical channels. The gas inlet channel includes a shut-off valve and a gas mass flow controller from Brooks. The following operating modes are provided:

- two independent gas supply channels, each of which supports a gas flow set for it;
- one channel operates in a fixed gas flow rate mode, the other - in the chamber pressure stabilization mode;
- both channels operate in the pressure stabilization mode in the chamber, maintaining a predetermined ratio between the mass flow rates through the channels.

IOP Conf. Series: Materials Science and Engineering **781** (2020) 012014 doi:10.1088/1757-899X/781/1/012014

3. Electron-beam gun

The «ELTA-60» electron-beam gun manufactured by the Tecarte company is shown in Figure. 2. The electron source 1 is intended to create and form an electron beam. In the intermediate housing 2, which is the connecting link between the electron source and the focusing-deflecting system, a locking mechanism 4 is installed, which serves to cut off the volume of the electron source from the rest of the gun volume. Focusing-deflecting system 3 is designed to focus and deflect the electron beam during processing. The vacuum system is designed for differential pumping of the gun, and consists of an angular vacuum conduit with a flange for connecting a turbomolecular pump (TMP) 5. The gun cooling system is connected in series with the TMP cooling system. The vacuum system provides a pressure of no higher than 10^{-2} Pa in the electron source.

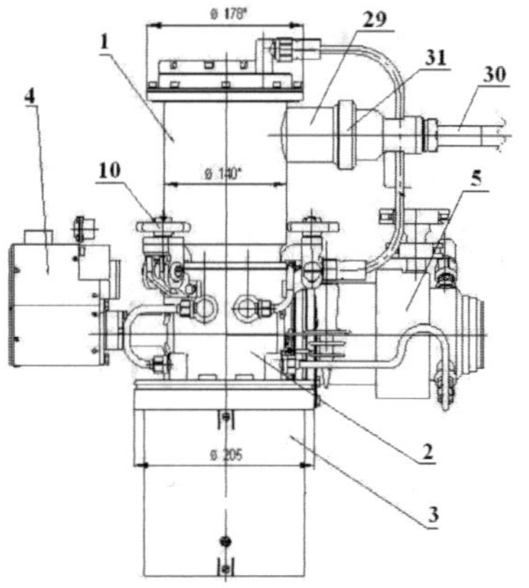

Figure 2. External design of the electron beam gun «ELTA-60»: 1 – electron source; 2 – intermediate housing; 3 – focusing-deflecting system with double refraction of the beam; 4 and 5 – vacuum system

The electrons emitted by the heater are accelerated by a voltage of the order of 1 kV and bombard the cathode, heating it. A lanthanum hexaboride tablet is used as an electron source.

Adjustment and stabilization of the electron beam current is carried out by changing the bias voltage at the control electrode. To focus and deflect the electron beam, a focusing lens and a deflecting birefringent system are used. The electron source is located vertically on the top plate of the chamber, the electron beam is directed down. The characteristics of the electron gun are given in Table 4.

Table.4 Characteristics of the electron gun

Parameter	Value
The value of the accelerating voltage, kV	60±1
Range of change of welding current, mA, not less than in the "welding" mode	1-250
Frequency modulation current, Hz	0 – 50
Maximum deviation angle, deg., not less than	±10
Maximum parallel electron beam transfer, mm, not less than	±12
Supply voltage, V	380
Supply voltage frequency, Hz	50
Maximum power consumed by the equipment, kW	50
The minimum range of variation of the bombardment current, mA	10-50
Water flow for cooling, m^3 / s	$3.58 \cdot 10^{-4}$

4. Workbench

The processed product is placed on a water-cooled workbench (Figure 3), mounted inside the chamber on its lower plate. The table is equipped with stepper motors 1, 2, and provides movement of the processed sample by 200 mm in a horizontal plane in two mutually perpendicular directions, if the

IOP Conf. Series: Materials Science and Engineering **781** (2020) 012014 doi:10.1088/1757-899X/781/1/012014

sample is fixed in rotating faceplates 3, then its rotation around an axis parallel to one of the axes of horizontal movement is ensured. To focus the electron beam on the treated surface, a vertical movement of the electron beam gun by 100 mm is provided. The selected solution allows electron beam processing of both flat and cylindrical surfaces of the products.

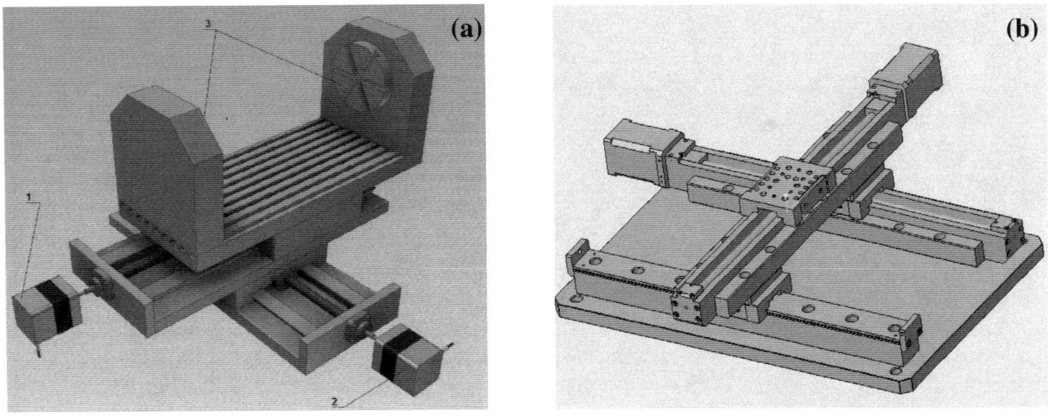

Figure 3. Workbench for placing samples (security elements are not shown)

The positioning system serves to change the position of the workpieces / products inside the vacuum chamber relative to the electron beam source, and includes a two-coordinate workbench, a spindle for rotating the part and a drive for moving the electron beam source relative to the part in the vertical axis in order to focus the electron beam.

The positioning system consists of four FL57STH stepper motors, four SMD-4.2 stepper motor drivers, driver power supplies and four EM253 expansion modules in the control controller. The controller sets the necessary speed, acceleration and braking times, searches for reference marks, monitors the status of the sensors of extreme positions. Communication between the controller and the stepper motor drivers is carried out via the "Step / Direction" interface.

5. Conclusions

The developed electron-beam machine «LUCH» implements various technologies for processing glass and ceramic materials: melting, welding, cutting, polishing and dimensional microprocessing. The most promising areas for research are dimensional microprocessing of LTCC ceramics. In addition, due to its characteristics, the installation has great potential for use in additive technologies.

References

[1] Węglowski M S, Błacha S and Phillips A 2016 *J. Vacuum* **130** pp 72-92
[2] Hellberg S et al 2018 *J. Phys.: Conf. Ser.* **1089**(1) p 012003
[3] Chen Y 2015 *J. Microelectronic Engineering* **135** pp 57-72
[4] Spinney P S et al 2010 *J. Nanotechnology* **21**(37) p 375301
[5] Kita J et al 2002 *J. Microelectronics International* **19**(3) pp 14-8
[6] Hagen G and Rebenklau L 2006 *Conf. 1st Electronic System integration Technology Conference* **1** pp 642-7
[7] Sebastian M T and Jantunen H 2008 *J. International Materials Reviews* **53**(2) pp 57-90
[8] Rebenklau L, Wolter K J and Hagen G 2006 *Int. 29th International Spring Seminar on Electronics Technology* pp 55-63
[9] Wang G et al 2006 *J. IEEE transactions on electronics packaging manufacturing* **29** pp 32-41

[10] Kita J et al 2002 *J. Microelectronics International* **19**(3) pp 14-8
[11] Nowak K M, Baker H J and Hall D R 2006 *J. Applied Physics A* **84**(3) pp 267-70
[12] Miyake H, Tanaka Y and Takada T 2007 *J. IEEE Transactions on Dielectrics and Electrical Insulation* **14**(2) pp 520-8.
[13] Rau E I et al 2013 *J. Russian Microelectronics* **42**(2) pp 89-98
[14] Burdovitsin V A, Klimov A S and Oks E M 2009 *J. Technical physics letters* **35**(6) pp 511-3
[15] Burdovitsin V A et al 2010 *J. Plasma Sources Science and Technology* **19**(5) p 055003.
[16] Okada A et al 2011 *J. CIRP annals* **60**(1) pp 575-8

Analysis of ion-plasma technology application in medical products proceedings to be used for maxillofacial surgery (Review)

A.V. Demiyanova*, E.A. Mitrofanov, S.B.Simakin**, A.M. Sipkin*****

*Moscow, State-Financed Health Care Facility of Moscow Region (SFHCFMR) "Moscow Region Scientific Research Clinical Institute named after M.F. Vladimirskiy" (MRSRCI), 129110, Shchepkin street, 61/2, SFHCFMR Odintsovo District Hospital №2 , **Moscow, Stock Company (SC) "Scientific Research Institute of Vacuum Technique named after S.A. Vekshinskiy" (SRIVT), 117105, Nagorniy proyezd, 7 Moscow, State-Financed Health Care Facility of Moscow Region (SFHCFMR) "Moscow Region Scientific Research Clinical Institute named after M.F. Vladimirskiy" (MRSRCI), 129110, Shchepkin street, 61/2

nastazzzy@mail.ru, aleksipkin@bk.ru, E.Mitrofanov@niivt.ru, S.Simakin@mail.ru

Abstract. This paper concerns with published data review about properties of materials and coatings, which are applied for maxillofacial miniplates manufacturing and coating. Such products are known to be widely used for osteosynthesis in maxillofacial surgery because of its considerable advantages over other forms of fixation. Most of alloys contain toxic metals. Application simultaneously at the same place materials, which possess different conductivity results on galvanic effects. It has been supposed to use deposited in vacuum silicon carbide (SiC) coating to decrease negative influence of toxic elements to a man. This coating successfully has been applied in dentistry.

1. Introduction

Nowadays all round the world it is being noted the tendency of injury level increasing. This phenomenon is result from development of industrialization of society, availability of different types of weapon, increased general level of a stress among the population of cities and other factors. [1,2] Among the total number of all kinds of injuries, injuries in the orofacial region bones range up about 16%.Among them mandibular injuries are as much as 80%, a zygomatic maxillary complex - 12%, injuries of maxilla and nose bones - about 8%.[2,3] Also, it should be noted that among the patients who are on hospital treatment of a dental profile, the total number of injuries in the maxillofacial region is about 40% and it does not tend to decrease. [4] Among patients who had injuries in the orofacial region and were operated in hospital department of maxillofacial surgery of the Odintsovo District hospital No. 2 during 2018, about 78% underwent osteosynthesis surgery of facial skeleton bones and about 40% from them had the burdened anamnesis.

Now the most frequent type of surgery at bone fractures in the oral and maxillofacial region is the osteosynthesis by means of miniplates and screws, which are made of titanic alloys. This type of operation is known to provide reliable fixation of fragments during the necessary period and their full immobilization. Examples of miniplates produced by firms of Great Britain are represented in the Fig. Two top plates were made by Champy of stainless steel (17.2 % Cr, 14 % Ni). Below there are plates produced by Champy, Leibinger, DePue and Stratec/AO firms. These four plates were made from titanic alloys. [5]

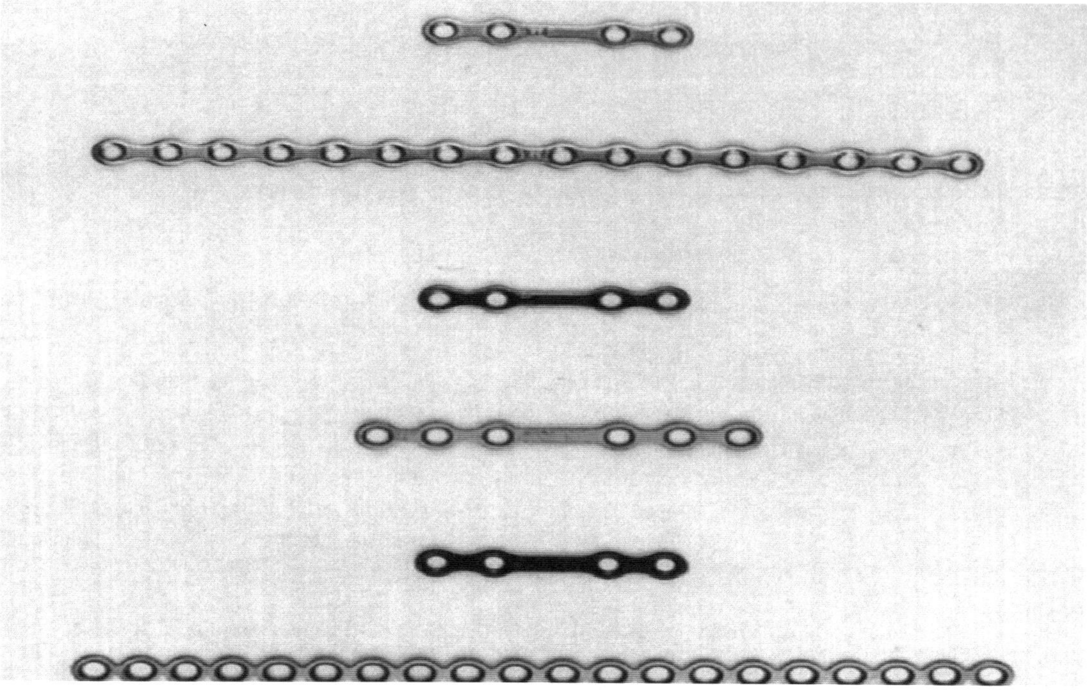

Figure.1 Photograph of the miniplates for facial fractures produced in Great Britain. They are known to be widely applied in maxillofacial surgery

For the first time this type of bone fractures treatment was applied by H. Hansmann and W. A. Lane. However, exactly V.V. Baryash was the first, who suggested to use titanic plates at treatment of bone fracture in oral and maxillofacial surgery. Pure titanium is not applied to produce miniplates for osteosyntethis because of its insufficient physical, mechanical and strength properties. Titanic alloys 6AL4V и 6AL4V ELI consist of titanium and the alloying additives: 6% of aluminium and 4% of vanadium. They are known to be used to produce plates, needles, pins, which are applied for osteosynthesis as they can maintain big loadings.[6] It is known an alloy titanium nikelid (NiTi). It was called nitinol. This material possesses to reconstruct primary form of a deformed detail after its heating ("thermomemory"). It contains 55% of nickel and 45% of titanium. It is known to be used in medicine, including in particular, for self-expanding stents manufacturing.

When using open methods of surgical treatment to operate fracture region there are applied such methods as skeletonization, reposition and fixation of bone fragments. However, it should be noted that surgical intervention enhances risk of secondary infection of a wound during the postoperative period. Nowadays the percent of complications during the postoperative period is considerable. [7] For example, the most widespread complication of inflammatory genesis is chronic traumatic osteomyelitis of the mandible, which is as much as from 26.3 up to 37.2 % of the total number of complications of facial bone fractures. [8]

Besides infectious complications, it is necessary to allocate such undesirable phenomenon as development of a metallosis and galvanic effects at persons who have a combination of different grades of steel and/or titanium in a plate and fasteners. [9]

It is also important to note, that there is a number of metals which availability in an organism adversely influence the rate of wound healing. Considering it, there are known alloy combinations to be excluded: chrome and cobalt, titanium and vanadium, nickel in high percentage concentration together with any metals. [9,10,11] Worse still, it should be noted that during long contact of person`s tissues and various biological environments with the plates made of the alloys containing the alloying additives, some plates and other medical products partially collapse and a part of chemicals, which are contained in alloy compositions, penetrate in surrounding tissues. As a result, the maintenance of Fe, Ni, Ti, Cr. increases in an organism. The excess maintenance of metal ions in an organism has toxic effect. The toxicity is a result from binding of "metal poisons" in an organism with functional groups of the vital substances that break normal functioning of cells in fabrics. [10]

Corrosion properties of plates and fasteners directly influence on pH of surrounding liquids that can result on development of purulent - inflammatory diseases after operational intervention. [9,10]

Thus, the problem to prevent development of secondary complications during the postoperative period at patients with bone injuries in the orofacial region is acute today

So, to prevent secondary infection, it is known to be applied antibacterial therapy during the postoperative period. However, this therapy cannot provide completely prevention of postoperative complication as antibiotics can impact only microflora. Therapy by antibiotics cannot prevent influence of the local and general factors on the probability of an infection complication development. [12] To prevent influence of local factors on development of secondary infection complications and to prevent an emission of toxic ions from medical products into tissues of a man, nowadays there are being investigated biologically active and inert materials as an alternative to products made from steel and titanic alloys. For example, the anode coating from titanium dioxide (TiO_2), which is formed by means of microarc galvanic oxygenating, exists and it is known to be applied in medicine. Also it is known the coating from titanium nitride (TiN) which is formed by method of arc deposition in vacuum. It is known to be applied for deposition, in particular, metallic dental prostheses. However, titanium dioxide (TiO_2) is found to be a genotoxic material. Traditional methods and the equipment for vacuum arc deposition demand high temperature of a sample surface, in particularly from 300 up to 850°C. It causes deformation of products, especially thin structural elements. [11,13,14,15]

The silicon carbide coating (SiC) does not contain any metal atoms. It has been successfully applied in stomatology to cover bases of removable dentures. [16,17] Application of a bioinert bioresistant coating made from silicon carbide to change properties of medical product surfaces, in particular, miniplates and fasteners made from titanic alloys, at surgical treatment of non-ballistic wounds of facial skull bones can become a possible version of the presented problems solution in maxillofacial surgery.

2. Conclusions

Summarizing the aforesaid, it is possible to assume that application of silicon carbide coating deposited in vacuum on external surfaces of the medical products, which are applied in maxillofacial surgery, in particular, miniplates and fixing screws will provide possibilities:

- to prevent penetration of toxic components from metal alloys and polymers in person`s tissues;
- to prevent galvanic effects in cases of direct contact and/or contact through conductive liquids of a person, in particular, blood, saliva, with plates and screws made from conductive materials which possess different conductivity;
- to prevent biodestruction of product materials;
- to reduce adhesion of microorganisms to medical product surfaces;
- to prevent biological fouling of product surfaces;
- to provide good adhesion of the coating to medical product surfaces.

References

[1] Abdulkerimov T.H., Kostina I.N. Dinamics of structure of oral and maxillofacial trauma during 10 years. In collection of scientific articles: Actual questions of modern medical science and public health service. Materials of the III International scientific-practical conference of young scientists and students, II Forum of medical and pharmaceutical universities of Russia “For quality education”. Yekaterinburg, 2018; p.17-21.

[2] Shargorodskiy A.G., Stefantsev N.M. Traumas of soft tissues and bones of face. Moscow, All-Russian training-scientific-methodological center, 2000, p.240;

[3] Bahteeva G.R., Kuzmin A.S. Statistical investigation of oral and maxillofacial traumas. Bulletin of Medical Internet Conferences (ISSN 2224-6150) 2012, 2 (11);

[4] Babkina T.M., Demidova E.A. Modern approaches in oral and maxillofacial trauma diagnostics. “Light of medicine and biology”. 2013,;4;7-11;

[5] Cawood J.I. Small plate osteosynthesis of mandibular fractures. British Journal of Oral and Maxillofacial Surgery. 1985; 23; 77-91;

[6] For which purpose titanium plates are used in case of fracture? http://bezperelomov.com/obshee/titanovaya-plastina-pri-perelomax.html;

[7] S.A. Safarov, A.E. Shcherbovskih, Yu.V. Petrov, I.M. Bayrikov.Clinical and functional rationale for the use of intraosseos composite materialscovered screw retainers for mandibular fracture osteosynthesis. Samara State Medical University, Samara, Russia; Kazan medical journal. 201495(2); 219-223;

[8] Mamitova A.B., Urgunaliev B.K., Moldaliev E.K. Traumatology of oral and maxillofacial region. Bishkek, 2014;

[9] What is galvanism in stomatology: therapy, symptoms, photographs. http://topdent.ru/articles/galvanism-v-polosti-rta/html;

[10] Removal of a plate, pins and screws after ankle fracture: operation and rehabilitation period. http://nogostop.ru/golenostop/lodyzhka/udalenie-plasttiny-posle-pereloma-lodyzhki.html.;

[11] Zaitseva A.G. Diagnostics and therapy of galvanism in mouth cavity. Extended abstract of dissertation. Russia, Saint Petersburg, 2005;

[12] Yu. P. Savchenko, V. V. Malishko, I. V. Golikov, S. N. Szczerba Causes of suppurative complications in the postoperative period and Prevention Department of general surgery Kuban state medical university, Russia. Kuban scientific medical journal. 2013; 7; 112-114;

[13] Grebenshchikova M.M., Nigmatullina R.A.,Haidarova L. M. Toxic investigation of materials for orthopaedy with ion-plasma coatings. Bulletin of Kazan Technological University, 2014, p 49 – 51;

[14] Korolev E. Titanic nanoparticles can initiate cancer!-Notes of provincial journalist. Forum- orbita96. http://forum.orbita96.ru/index.php/topic/181-nanochastitcy-titana-mog/;

[15] Abilev S.K., Glazer V.M. Mutagenesis and foundations of genotoxicology. Textbook. Moscow, Saint Petersburg, 2015;178-179.;

[16] Voronov I.A., Mitrofanov E. A., Kalinin A.L., Simakin S.B., Didenko L.V., Avtandilov G.A. Development of new coating from silicon carbide to protect dental prostheses from biodegradation. Journal of Russian dentistry. 2014;1;4-8.

[17] Certificate of state registration of medical product.25.10.2017. №P3H 2017/6407. Medical product: Stomatological material: Silicon carbide coating to protect denture bases “Armor” in accordance to Technical Specification 9391-001-07604422-2015.

IOP Conf. Series: Materials Science and Engineering **781** (2020) 012016 doi:10.1088/1757-899X/781/1/012016

Neural Network modeling of Thin Films Deposition Processes in the Master's Degree Programs "Electronics and Nanoelectronics" and "Nanoengineerig"

Ekaterina V. Panfilova, Elena N. Galaganova

Bauman Moscow State Technical University, Moscow, Russia

panfilova.e.v@bmstu.ru

Abstract. In this paper we consider the application of artificial neural networks in the Master's degree programs. Nowadays neural network models have become a powerful tool of scientific research for engineers and students. Besides successful research depends on confided usage of common methods and algorithms of statistical community special attention in degree programs is given to data analysis and modeling methods. Students study modern neural networks software products, methods of data preparing, designing and training neural network and then apply neural networks algorithms in practice. The neural networks modeling plays a significant role in students education. During it students learn to create, tune and train their own neural networks and get full and detailed understanding of research in computer vision. Neural Networks models are successfully presented in graduation works. In this paper a few examples of how neural networks are used in thin films deposition processes modeling are presented.

1 Introduction

Successful economic development is impossible without nanotechnology. Nowadays there is an increasing demand worldwide for specialists knowledgeable in nanotechnology (Panfilov, Kolesnik, Ryabov & Sidorova, 2017). In conformity on modern requirements of the third generation standard in highest professional education and Master's Degree Programs "Electronics and Nanoelectronics" and "Nanoengineerig" (Bauman Moscow State Technical University, Educational plans and programs, 2019) student's scientific work is an obligatory part of education in the Chair of Electronic Technologies in Mechanical Engineering of Bauman Moscow State Technical University. For four semesters students of Masters's degree studies under the guidance of scientific supervisor work on a specific scientific problem such as investigation and development of modern technologies, engineering of technical processes and equipment. Students can, depending on their background and interests, choose the research fields such as processing techniques, specialized equipment and key materials of electronics, nanoelectronics and nanoengineering: vacuum technics (Vasilev, Malevannaya & Moiseev, 2017), vacuum (Kolesnik, Zhuleva, Predtechenskiy, Kyaw & Phyo, 2017) and colloidal (Kuleshova, Panfilova & Prohorov, 2018) thin films deposition, various lithographic and etching processes, etc.

The nature of nanoscience is cross-disciplinary and students specialize not only in nanoelectronics and nanoengineering, but in chemistry, nanophysics, quantum mechanics, etc. Besides successful research depends on confided usage of common methods, algorithms and tools of statistical community special attention in degree programs is given to modern, current and demanded data analysis and modeling methods. Master's degree program include the course "Neural Network modeling of Complex Technical Systems". It is taught in the third semester of the Master's programs "Electronics and Nanoelectronics" and "Nanoengineerig".

2 Methodology of Education

Neural networks can be used in technological process control, code sequence prediction, failure analysis, machine vision, signal identification, nonlinear modeling, etc (Moroz, Cheremisin, Cherkashina & Kholod, 2016; Rene, López, Kim & Park, 2013; Huo, He, Su, Xi, & Zhu, 2013). A neural network is a computing model whose layered structure copies the structure of neurons in the brain, with layers of several connected single processing elements, called neurons. The neurons are connected by weights. These weights are tuning in during learning or training until the error becomes minimum. Neural networks are training over many examples. Training data is a specific number of observations, for each of which values of several variables are indicated. Most of these variables are set as inputs and the network will learn to find a match between the values of the input and known output variables. To train the artificial neural network, data is required to be divided into three datasets as training, validation and test data. The mathematics of neural networks is well developed and many computer programs are available to create, train and exploit neural networks.

Every neural network application is special, but developing the network follows these steps:

- preparing the data,
- creating the neural network,
- training the neural network,
- validating the network's results,
- tuning the neural network,
- integrating the neural network into a project.

According to the steps of networks development the course consists of five parts. The first part of the course discusses applications of neural networks to practical engineering problems in technological and analytical equipment, sensor and control systems, pattern recognition, clustering. The second part of the course is devoted to accessing and preparing the data. The third part of the course gives an introduction to basic artificial neural network architectures. It provides a clear and detailed survey of fundamental neural network types - multilayer perception, radial basis function network, probability neural networks and Kohonen neural network (Gurney, 2014). Students learn how to construct a set of simple elements - artificial neurons. The fourth part of the course discusses training of neural networks, optimisation algorithms and learning rules: backpropagation and several variations of backpropagation, such as the conjugate gradient, Levenberg-Marquardt algorithm, gradient descent and gradient descent with adaptive learning rate (Sapna, Tamilarasi & Kumar, 2012) . At the fifth part of the course students learn how to validate the network's results and final-tune network parameters. Figure 1 shows the course structure.

Seminars are devoted to the software products: Statistica Neural Networks (Statistica Automated Neural Networks, 2019), Neural Network Toolbox for MATLAB (now - Deep Learning Toolbox) (Deep Learning Toolbox, 2019) and NeuroShell (NeuroShell 2, 2019). During the homework students learn to create, tune and train their own neural networks and get full and detailed understanding of research in computer vision.

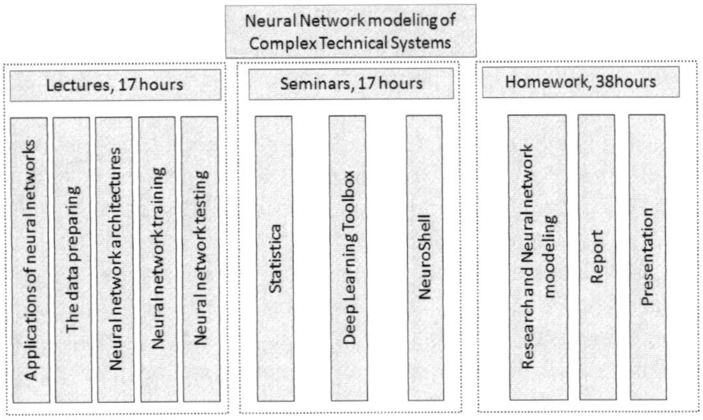

Fig. 1. "Neural Network modeling of Complex Technical Systems" course structure.

After completing the study of the discipline "Neural Network modeling of Complex Technical Systems" the student are expected to:

- understand neural networks algorithms;
- know modern neural networks software products;
- be capable to prepare data;
- be capable to design, train, tune and test neural network;
- be capable to apply neural networks algorithms in practice.

3 The Results of Education and Discussion

Since the course has rather practical than theoretical nature, it plays an important role in student's research work and development of their graduation works. It is advised to use the homework for solving scientific technical tasks in individual research projects.

There are a few examples of how neural networks are used in students projects in this article:

- modeling of the metal islands thin films growth while vacuum evaporation process using multilayer perception,
- forecasting the properties of photonic band colloidal thin films using radial basic function neural network.

Regression models (Dreiseitl & Ohno-Machado, 2002) are used in students research most often. They describe the relationship between a response (output) variable and predictor (input) variables of technological process, equipment, product or material. Here are a few examples from homeworks and graduation works.

Typical homework report includes abstract, introduction, description part, designing part, results of neural network modeling, discussion and conclusion. Investigating system or process, its input and output parameters and available statistical data are presented and analyzed in description part. The designing part contains the rationale for choosing inputs and outputs, the type of neural network, transfer and error functions and learning algorithm for network training. Description of neural network training, validating and application are presented as a results of modeling. New understanding of the problem, interpreting the findings and the prospects of using modeling results are presented in the discussion section. In conclusion section the arguments involved in the body paragraphs are summarized. The results of neural networks modeling are showed not only in homework reports but special time is reserved for students presentation. Thus homework's neural networks modeling is a completed work. It plays a significant role in student's research and therefore homework reports are widely used as a part of grading works.

Figure 2 illustrates the modeling of colloidal photonic crystal (Armstrong & O'Dwyer, 2015) film deposition process. The films were deposited by vertical lifting method. It has gained great interest because of its simplicity and ability to vary the conditions.To predict the wavelength and reflectance at photonic band gap (PBG) of film the following initial data was necessary: colloidal particles size and material, colloidal solution concentration and lifting velocity. All data were determined directly from the results of the series of reflectance measurements of the polystyrene and silica particles films at various velocities (Kuleshova etal., 2018). The obtained reflectance spectrum shows that the wavelength and reflectance of PBG are in dependence on the lifting velocity, colloidal concentration and particles size for both materials. But self-assembly processes that occur during vertical deposition have been characterized by uncontrolled formation of defects over larger length scales.

The radial basic function (RBF) neural network was used in the simulation as the most precise for the task of predicting nonlinear relation. The RBF neural network model with 4 input neurons, one hidden layer with 20 neurons and 2 output neurons has been created. It describes the relationship between the wavelength and reflectance at PBG (output variable) and colloidal particles size and material, colloidal solution concentration and lifting velocity (input variables).

Neural network was used to optimize the effect of the above-mentioned factors. The revealed dependences allowed to determine the optimal conditions for obtaining high photonic crystalline quality of opal film: the highest values of the relative area of ordered structure and reflectance at the PBG. Obtained at optimal conditions opal film has closest packed structure consisting of hexagonal close packed layers and high reflectance. The results of modeling can be used for various applications of photonic crystal films: nanophotonics, laser technics, luminescence microscopy of biological objects, plasmonics, sensory, etc.

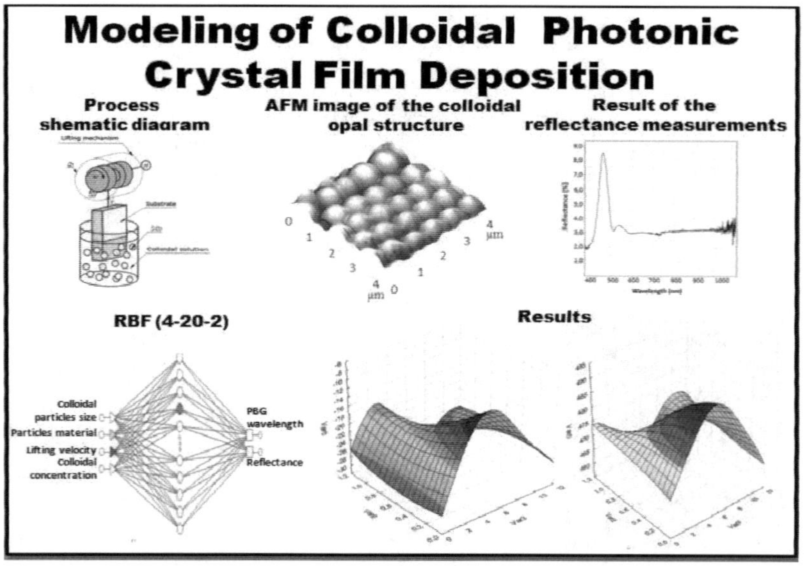

Fig. 2. Modeling of the colloidal photonic crystal film deposition by vertical lifting process.

Figure 3 shows an example of modeling of metal islands film growth (Sidorova, Pronin & Isaeva, 2018). Multilayer perception (MLP) with two hidden layers was used in the simulation because of it extrapolation ability (Hettiarachchi & Minns, 2005). The best way to determine the number of hidden neurons is to train several networks. Preliminary option analysis included two MLP and two RBF neural networks. It was revealed that a few number of hidden neurons result in a higher training error and too many hidden neurons results in a higher validation and testing error.

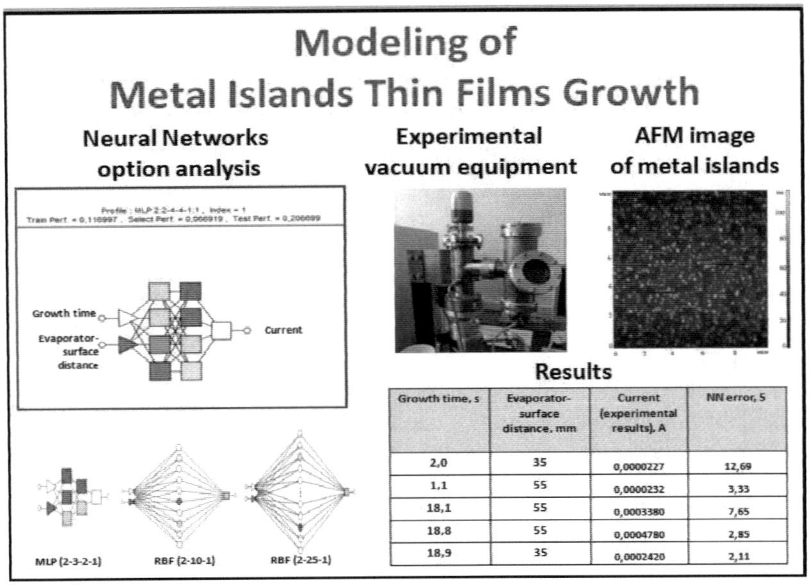

Fig. 3. Modeling of the metal islands thin films growth while vacuum evaporation process.

The modeling approach described in the course "Neural Network modeling of Complex Technical Systems" allows to get useful information for effective student's scientific investigation. Neural network models help students understand the thin films deposition processes dependencies.

4 Conclusion

The principles of artificial intelligence is applied with the help of neural networks (Rene & Saidutta, 2008; Russell & Norvig, 2016). For engineers and nanoscientists neural network models have become a powerful tool of scientific research and machine learning. The concept of neural network modeling has widespread applications in the field of applied science and engineering, in particular, in nanotechnology. The minimum value of errors, the high level of objectivity in the results of output predicting and real-time control are the advantages of neural networks in comparison with other methods.

The results acquired with neural network models for the experimental study of nanotechnological processes show that the application of neural networks is sufficient. Neural Networks models are successfully used in students scientific and research projects and presented in Master's and Specialist's graduation works. The aim of the course is not only to introduce students to the neural network modeling, and also to develop self-learning abilities (Siriwongs, 2015; Cheng & Chau, 2013).

References

[1] Armstrong, E., & O'Dwyer, C. (2015). Artificial opal photonic crystals and inverse opal structures–fundamentals and pplications from optics to energy storage. *Journal of Materials Chemistry C*, 3(24), 6109-6143.

[2] Bauman Moscow State Technical University, Educational plans and programs (2019). Retrieved from http://www.bmstu.ru/abitur/studies-life/edu-programs/#eduprograms.

[3] Cheng, G., & Chau, J. (2013). Exploring the relationship between students' self-regulated learning

ability and their ePortfolio achievement. *The Internet and Higher Education*, 17, 9-15.

[4] Deep Learning Toolbox, 2019. Retrieved from: https://www.mathworks.com/products/deep-learning.html.

[5] Dreiseitl, S., & Ohno-Machado, L. (2002). Logistic regression and artificial neural network classification models: a methodology review. *Journal of biomedical informatics*, 35(5-6), 352-359.

[6] Gurney, K. (2014). *An introduction to neural networks*. CRC press.

[7] Hettiarachchi, P., Hall, M. J., & Minns, A. W. (2005). The extrapolation of artificial neural networks for the modelling of rainfall—runoff relationships. Journal of Hydroinformatics, 7(4), 291-296.

[8] Huo, S., He, Z., Su, J., Xi, B., & Zhu, C. (2013). Using artificial neural network models for eutrophication prediction. *Procedia Environmental Sciences*, 18, 310-316.

[9] Kuleshova, V. L., Panfilova, E. V., & Prohorov, E. P. (2018, September). Automated Device for Vertical Deposition of Colloidal Opal Films. In *2018 International Russian Automation Conference (RusAutoCon)* (pp. 1-5). IEEE.

[10] Kolesnik, L. L., Zhuleva, T. S., Predtechenskiy, P. O., Kyaw Hlaing, M., & Phyo Aung, Z. (2017, July). Processing of metallization technology aluminum oxide ceramics for electro-vacuum devices elements and power electronics devices. In *Journal of Physics Conference Series* (Vol. 872, No. 1).

[11] Moroz, A. N., Cheremisin, N. M., Cherkashina, V. V., & Kholod, A. V. (2016). Neural network modeling in problems of electrical grids modes prediction. *Electrotechnics and electromechanics*, (1).

[12] NeuroShell 2, 2019. Retrieved from: http://www.neuroproject.ru/aboutproduct.php?info=ns2info.

[13] Panfilov, Y. V., Kolesnik, L. L., Ryabov, V. T., & Sidorova, S. V. (2017, July). Research and development complex with remote access. In *Journal of Physics*: Conference Series (Vol. 872, No. 1, p. 012010). IOP Publishing.

[14] Rene, E. R., López, M. E., Kim, J. H., & Park, H. S. (2013). Back propagation neural network model for predicting the performance of immobilized cell biofilters handling gas-phase hydrogen sulphide and ammonia. *BioMed research international*, 2013.

[15] Rene, E. R., & Saidutta, M. B. (2008). Prediction of Water Quality Indices by Regression Analysis and Artificial Neural Networks. *International Journal of Environmental Research*, 2(2).

[16] Russell, S. J., & Norvig, P. (2016). *Artificial intelligence: a modern approach.* Malaysia; Pearson Education Limited,.

[17] Sapna, S., Tamilarasi, A., & Kumar, M. P. (2012). Backpropagation learning algorithm based on Levenberg Marquardt Algorithm. *Comp Sci Inform Technol (CS and IT)*, 2, 393-398.

[18] Sidorova, S. V., Pronin, M. A., & Isaeva, A. A. (2018, September). Automated Unit for Control of Initial Stages of Metal Islands Thin Films and Nanostructures Growth. In *2018 International Russian Automation Conference (RusAutoCon)* (pp. 1-4). IEEE.

[19] Siriwongs, P. (2015). Developing students' learning ability by dint of self-directed learning. *Procedia-Social and Behavioral Sciences*, 197, 2074-2079.

[20] Statistica Automated Neural Networks, 2019. Retrieved from: http://statsoft.ru/products/Statistica_Neural_Networks/.

[21] Vasilev, D. D., Malevannaya, E. I., & Moiseev, K. M. (2017, July). Vacuum coating system for deposition of superconducting W x Si (1–x) ultrathin films used in single photon detectors. In *Journal of Physics*: Conference Series (Vol. 872, No. 1, p. 012027). IOP Publishing.

IOP Conf. Series: Materials Science and Engineering **781** (2020) 012017 doi:10.1088/1757-899X/781/1/012017

Methods for improving the structural 3D ordering of self – assembled photonic crystal coatings obtained by electrophoresis deposition

E V Panfilova[1] **and D A Ezenkova**[1]

[1] Bauman Moscow State Technical University, 105005, Moscow, Russia

E-mail: panfilova.e.v@bmstu.ru

Abstract. The article describes the main properties of self – assembled colloidal photonic crystal coatings. A comparison of coating deposition methods is presented. Electrophoresis deposition method has gained great interest because of its possibility of conditions variation.The main factors affecting the quality of the colloidal opal films during electrophoresis deposition are studied. The results of experimental researches on increasing the ordering of photonic crystal coatings are presented. It was found that the voltage range of 0.7 - 2.7 V and pH of 7 allows one to obtain thick films with a pronounced PBG.

1. Introduction

Self-assembled photonic crystal (PhC) coatings are a class of materials that are most promising for modern science and technology. Interest in these coatings is due to the refractive index periodic changes in one, two or three dimensions and a photonic band gap (PBG). PhC films and coatings are widely used in electronics [1], optoelectronics [2], power electronics [3], solar power [4], spintronics [5], medicine [6], military and aerospace industries [7] and power engineering (supercapacitors) [8] on their own and as a part of multilayer structures. The structure of opal PhC is a matrix with highly ordered three dimensionally periodic cubic lattice of closely packed microspheres with a equal diameter of 50 … 1000 nm [9]. Self-assembly of polymer or silica (SiO_2) colloidal particles is one of the most simple and low-cost methods for the formation of photonic crystals (PhCs) [10]. Opal matrices are used to obtain PhCs, inverse structures and 3D nanocomposites as it shown in figure 1.

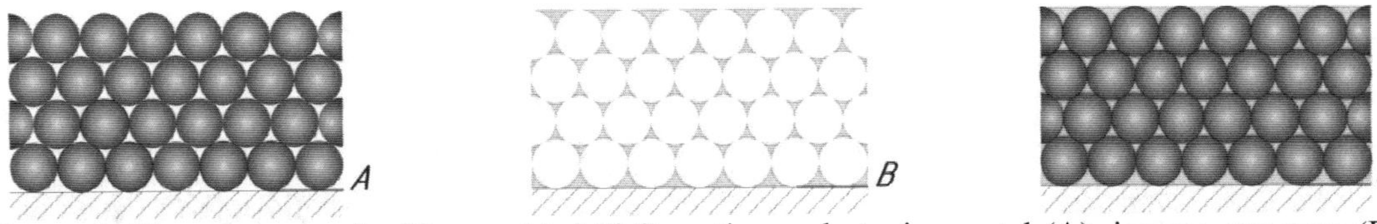

Figure 1. The structure of self-assembled PhC coatings: photonic crystal (A), inverse structure (B), 3D nanocomposite (C)

Interest in 3D – nanocomposites is due to a combination of ordering and matrix geometry, as well as properties of the implementation material [11]. By varying the diameter of the matrix microspheres one can change the voids size (10 … 600 nm), which in turn affects the properties of the implementation material. Due to the ordering of the structure, it is possible to create periodic nanoparticles arrays.

While various strategies have been developed to direct the self-assembly of colloidal particles, fabrication of 3D ordering transferrable colloidal coating with controllable PhC structure still remains a major challenge. In this study we determine the electrophoresis deposition process modes to make possible deposition of qualitative PhC coating.

2. Method description

At that moment, there are a number of widely used methods [10,12,13] for creating self –assembled PhC coatings. These technologies are presented in the table 1.

Table1. PhC coatings forming methods

Method	Process time	Film thickness	Nanocomposite forming	Disadvantages
Sedimentation	Weeks, month	Not limited	2 stage process	Long process time
Vertical deposition	Hours	<10 layers	2 stage process	Low repeatability
Electrophoresis deposition	Minutes, hours	up to hundred μm	Simultaneous deposition	Particle surface must be charged

It was revealed that the electrophoresis deposition method allow to form well-ordered structure in a wide range of thickness at high speed, and this method is also suitable for the formation of 3D – nanocomposites in one iteration. The essence of this method is described below.

The electrophoresis deposition method combines the self –assembly of microspheres from a colloidal system under the action of gravity with the application of external electric fields. The scheme of the process is shown in figure 2. The advantageof this method is a wide range of variation parameters, which makes it possible to obtain highly ordered coatings. The main parameter is potential difference supplied to the interelectrode space. It allows varying the microspheres deposition rate. Matrix formation is based on the self-assembly in an open dissipative system. The main driving force of the process is the evaporation of the solvent and, as a consequence, the change in the volume of the system. The main driving force depends on the temperature and vapor pressure of the solvent in the environment. The application of external electric fields allows to control and influence self-assembly process. Thus we get better quality of the coating structure.

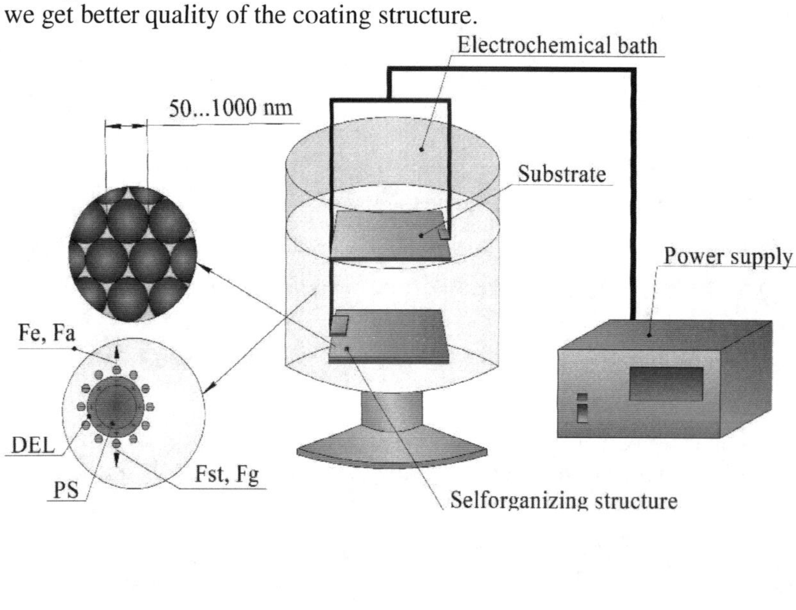

Figure 2. The schematic diagram of the electrophoresis deposition.

The process of electrophoresis deposition is possible due to the presence of a double electric layer (DEL) of microspheres. DEL causes uncontrolled behavior of microspheres during the deposition process. The density of DEL is determined by the value of ζ - potential, which can be set by changing the pH of the colloidal system as it shown in figure 3.

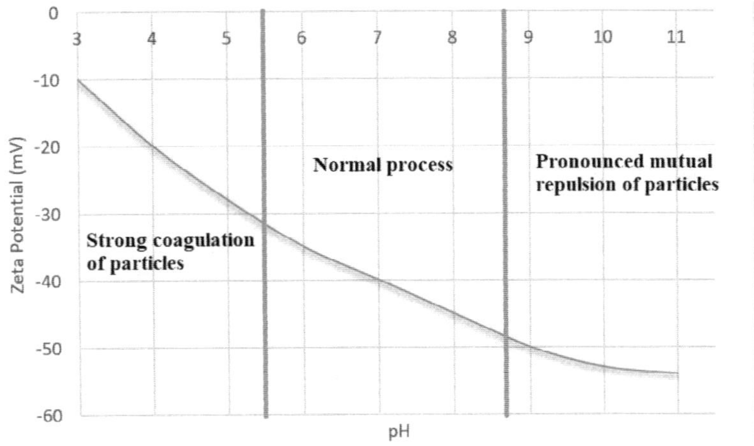

Figure 3. Microspheres ζ – potential vs pH of the colloidal system.

However, if pH significantly different from neutral, the formation of an ordered matrix frame is difficult. For a pH range of 3 ... 5, the value of ζ - potential and the density of DEL become insufficient to prevent coagulation of particles in solution. At pH = 9 ... 11, the formation of an ordered coating is difficult, due to the presence of repulsive forces, therefore, pH = 7 was chosen for further studies [14].

The study was carried out on copper substrates. In order to remove fats, dust and oxides from the surface of the substrate before the experiment they were purified in acetone for 2 min, isopropyl alcohol - for 2 min. In order to remove products from the previous purification stage - in deionized water for 30 s.

The photonic properties of the obtained coatings were monitored as an output parameter by measuring the reflection coefficient of PBG by means Epsilon spectrophotometer (IZOVAC, Belarus).

3. Results and discussion

Colloidal polystyrene monodisperse latex and silica SiO_2 particles with a microsphere diameter of 520 nm were used as a material for the formation of opal matrix. A preliminary experiment was performed using pH of 7, the applied potential difference of 0.5 V, colloids concentration of 2 %, interelectrode distance of 15 mm and the deposition time of 5 min. 10% colloidal solution with particles diameter 220 nm at a lifting velocities of 1 mm/min and 0.8 mm/min. Figure 4 presents the reflection spectrum of the polystyrene and silica structures obtained during the preliminary experiment.

The similarity of the reflection spectra between the samples indicates a similar behavior of the silica and polystyrene microspheres under electrophoresis conditions. The structures were defected and many empty plots of substrate were seen. It was revealed that the use of polystyrene particles as a matrix material increases the reflection coefficient, which indicates an increasing of the structures ordering. This can be justified by the small heterogeneity of the sizes of polystyrene microspheres (~ 5%), while the heterogeneity of silica microspheres can reach 30%.

The film's areal density is proportional to the deposition velocity. In order to increase it a voltage range from 0.3 V to 2.7 V was chosen.

IOP Conf. Series: Materials Science and Engineering **781** (2020) 012017 doi:10.1088/1757-899X/781/1/012017

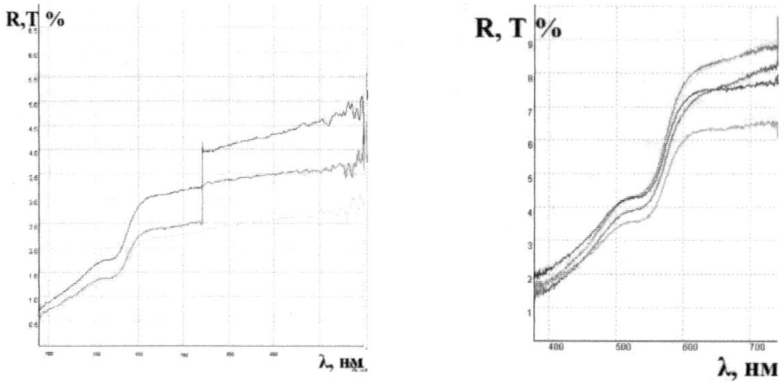

Figure 4. Spectrophotometric analysis of samples of silica (left) and polystyrene (right), obtained under certain conditions. For all samples reflection spectra obtained at three points on the surface of the substrate were obtained.

It was revealed that the deposited films were formed mainly on the anode and they covered the entire surface of the substrate. As a consequence of obtained periodicity obtained films showed opalescence: reflected colors that came from the Bragg diffractions in the periodic distribution of microspheres. Spectrophotometric analysis showed that all structures possessed a PBG in the wavelength range of 460 - 480 nm. A color-shift from green to red was found. The reflectance spectrums of such structures demonstrate a defined Bragg stop band at every sample as shown in figure 5. A similar position of the PBG indicates the identical behavior of microspheres under electrophoresis at the different values of the applied voltage. However, obtained in the voltage range of 2.7… 0.9 V coatings were characterized by brittleness, poor adhesion to the substrate surface due to significant residual stresses in the structure and appearance of large clusters of particles on the substrate.

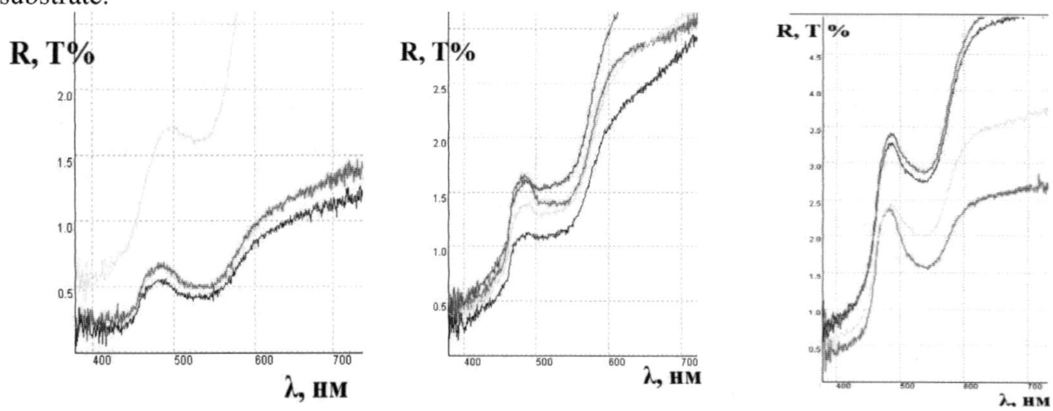

Figure 5. Reflection spectrum for samples obtained with a potential difference of 2.7 V, 0.9 and 0.3 V. For all samples reflection spectra obtained at three points on the surface of the substrate were obtained.

In order to eliminate agglomeration of particles in the solution we decreased voltage from to 0.3 V. As a result an increase in the reflection coefficient was noted, which may also indicate an increase in the ordering of the structure. No defects and particles clusters on the structures surface were noted. The coating obtained at a potential difference of 0.3 V possessed a pronounced at a wavelength of 480 nm and good adhesion to the substrate.

4. Conclusions

In the course of the study it was found that the voltage range of 0.7 - 2.7 V allows one to obtain thick films with a pronounced PBG in the wavelength range of 460 - 480 nm. However, the films obtained in this range possessed significant residual stresses, which led to the destruction of the structure. The voltage range of 0.3 - 0.6 V allows one to obtain well-ordered structures with a pronounced PBG in the wavelength range of 460 - 480 nm and the highest values of the reflection coefficient, as well as sufficient adhesion to the surface of the substrate. The authors see the development of the work in the research of the phenomenon of self-assembly in colloids.

References

[1.] Syritskii, A. B., & Panfilova, E. V. (2018, November). Investigation of opal nanostructures using scanning probe microscopy. In *IOP Conference Series: Materials Science and Engineering* (Vol. 443, No. 1, p. 012035). IOP Publishing.

[2.] Katyba, G. M., Zaytsev, K. I., Chernomyrdin, N. V., Shikunova, I. A., Komandin, G. A., Anzin, V. B., ... & Reshetov, I. V. (2018). Sapphire photonic crystal waveguides for terahertz sensing in aggressive environments. *Advanced Optical Materials*, *6*(22), 1800573.

[3.] DEDE, E. M., & Joshi, S. N. (2019). *U.S. Patent Application No. 10/347,601.*

[4.] Parchine, M., Kohoutek, T., Bardosova, M., & Pemble, M. E. (2018). Large area colloidal photonic crystals for light trapping in flexible organic photovoltaic modules applied using a roll-to-roll Langmuir-Blodgett method. *Solar Energy Materials and Solar Cells*, *185*, 158-165

[5.] Rinkevich, A. B., Samoilovich, M. I., Klescheva, S. M., Perov, D. V., Burkhanov, A. M., & Kuznetsov, E. A. (2013). Millimeter-wave properties and structure of gradient Co–Ir films deposited on opal matrix. *IEEE transactions on nanotechnology*, *13*(1), 3-9.

[6.] Yu, J., Lei, J., Wang, L., Zhang, J., & Liu, Y. (2018). TiO2 inverse opal photonic crystals: Synthesis, modification, and applications-A review. *Journal of Alloys and Compounds*, *769*, 740-757.

[7.] Gorelik, V. S., & Friman, A. V. (2013). Dispersion of electromagnetic waves in a resonant opal photonic crystal filled with Al 2 O 3: Cr 3+. *Inorganic Materials*, *49*(6), 577-580.

[8.] Kim, J. H., Kang, S. H., Zhu, K., Kim, J. Y., Neale, N. R., & Frank, A. J. (2011). Ni–NiO core–shell inverse opal electrodes for supercapacitors. *Chemical Communications*, *47*(18), 5214-5216.

[9.] Gorelik, V. S. (2008). Optics of globular photonic crystals. *Laser Physics*, *18*(12), 1479.

[10.] Kuleshova, V. L., Panfilova, E. V., & Prohorov, E. P. (2018, September). Automated device for vertical deposition of colloidal opal films. In *2018 International Russian Automation Conference (RusAutoCon)* (pp. 1-5). IEEE.

[11.] Gorelik, V. S., Kudryavtseva, A. D., Tareeva, M. V., & Tcherniega, N. V. (2014). Stimulated light scattering in synthetic opal filled with dielectrics. *Inorganic Materials*, *50*(12), 1217-1221.

[12.] Bartlett, P. N. (2004). Electrodeposition of Nanostructured Films Using Seff-Organizing Templates. *Interface-Electrophoresis Society*, *13*(4), 28-33.

[13.] Hatton, B., Mishchenko, L., Davis, S., Sandhage, K. H., & Aizenberg, J. (2010). Assembly of large-area, highly ordered, crack-free inverse opal films. *Proceedings of the National Academy of Sciences*, *107*(23), 10354-10359.

[14.] Palmer, T. R. (2009). *Investigation of electrophoretic deposition as a fabrication technique for high performance composites* (Doctoral dissertation, Massachusetts Institute of Technology).

The study of heat transfer by radiation when heating a substrate using an infrared heater with a reflector in a working chamber with cold walls

S P Bychkov and E V Zikyi

Moscow, Bauman Moscow State Technical University, 2nd Baumanskaya str., 5, building 1

E-mail: bychkov@bmstu.ru

Abstract. A model is constructed, based on the algebraic method for calculating heat transfer by radiation, to determine the maximum temperature of the substrate heating depending on the geometric and thermophysical parameters of the surfaces for the "heater-substrate-internal surface of the working chamber-working and reverse surfaces of the reflector" system. The results are verified experimentally.

The task is to determine the maximum heating temperature [1, 2] of the product at a given heater power [3] with known geometric and thermophysical parameters for the considered system of bodies. The base computational scheme is a system of 5 surfaces of the following type: 1 - heater, 2 - substrate, 3 - working chamber surface, 4 - working surface of the reflector, 5 - reverse surface of the reflector. The scheme is illustrated in Figure 1.

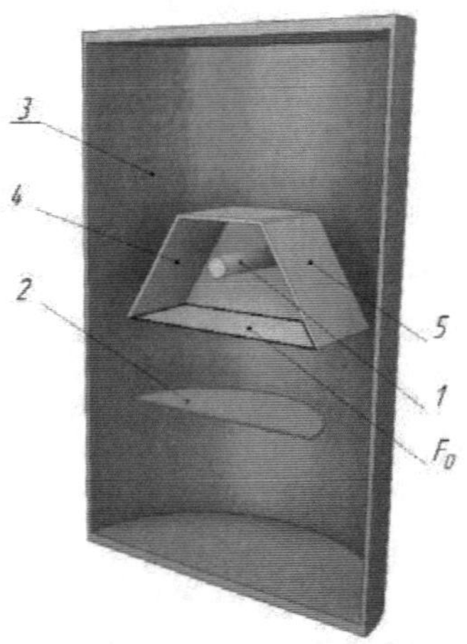

Figure 1. The computational scheme for determining the surfaces radiation fluxes: 1 – heater, 2 – substrate, 3 – working chamber surface, 4 – working surface of the reflector, 5 – reverse surface of the reflector, F_0 – auxiliary surface.

The initial system of equations for the resulting radiation fluxes Q_i in a system of 5 surfaces for the given conditions, obtained on the basis of the balance method, takes the form

$$Q_{12} + Q_{13} + Q_{14} = Q_1$$

$$(Q_{23} + Q_{24}) - (Q_{12} + Q_{32} + Q_{42}) = 0$$

$$(Q_{32} + Q_{34} + Q_{35}) - (Q_{13} + Q_{23} + Q_{43} + Q_{53}) = Q_3 \quad (1)$$
$$(Q_{42} + Q_{43}) - (Q_{14} + Q_{24} + Q_{34}) = Q_4$$
$$Q_{53} - Q_{35} = Q_5$$

and the system of equations for the remaining set of the significant view factors

$$\square_{12} + \varphi_{13} + \varphi_{14} = 1{,}0$$
$$\square_{23} + \varphi_{24} = 1{,}0$$
$$\square_{32} + \varphi_{33} + \varphi_{34} + \varphi_{35} = 1{,}0 \quad (2)$$
$$\square_{42} + \varphi_{43} + \varphi_{44} = 1{,}0$$
$$\square_{53} = 1{,}0$$

Here Q_{ij} is the part of the effective radiation flux of the i-th surface which falls onto the j-th surface:

$$Q_{ij} = Q_{effi} \cdot \varphi_{ij} = (E_{0i} \cdot F_i - R_i \cdot Q_i) \cdot \varphi_{ij} \quad (3),$$

where Q_{effi} is the effective radiation flux of the i-th surface in all directions; φ_{ij} – the view factor of the i-th surface onto the j-th surface; R_i – relative coefficient of reflection of the i-th surface

$$R_i = \frac{1 - \varepsilon_i}{\varepsilon_i} \quad (4),$$

where ε_i – integral emissivity coefficient of the i-th surface; E_{0i} – integral flux of own radiation of a black body at the temperature T_i of the i-th surface which is determined in accordance with the Stefan-Boltzmann law

$$E_{0i} = \sigma_0 \cdot T_i^4 \quad (5),$$

where F_i – value of the i-th surface.

Let us consider the model with the diffuse nature of the reflection of all surfaces in the case of cooling the surface of the reflector and the surface of the working chamber with water. The need for such cooling generally derives from two main objectives: maintaining the reflectivity of the surface during its operation and reducing undesirable gas evolution which increases with the temperature increasing [4, 5]. A distinctive feature of the water-cooled surface is the constancy of its temperature during the heating device operation. It is recommended not to exceed the level of 70-80°C.

In the presence of water cooling, the surface temperatures are considered constant

$$T_3 = T_{ch} = const, T_4 = T_5 = T_{cool} = const$$

which, upon transition to the fluxes of intrinsic radiation, means

$$E_{03} = E_{ch} = \sigma_0 \cdot T_{ch}^4, \qquad E_{04} = E_{05} = E_{cool} = \sigma_0 \cdot T_{cool}^4$$

With effective cooling, the temperatures of the working chamber and the reflector can be taken equal, while with diffuse reflection there is an equality of mutual surfaces of irradiation, the system of equations for determining the resulting fluxes has the form

$$\begin{aligned} \{\square_1 E_{01} F_1 &= p_h\, E_{02} F_2 + \varphi_{32} R_3 Q_3 + \varphi_{42} R_4 Q_4 \\ &= \varphi_{12} p_h + E_{cool} F_2\, \square_{23} E_{02} F_2 + \left(1 + (\varphi_{32} + \varphi_{34} + \varphi_{35}) R_3\right) Q_3 \\ &- \varphi_{43} R_4 Q_4 - R_5 Q_5 \\ &= -\varphi_{13} p_h + \varphi_{23} E_{cool} F_2\, \square_{24} E_{02} F_2 - \varphi_{34} R_3 Q_3 \\ &+ \left(1 + (\varphi_{42} + \varphi_{43}) R_4\right) Q_4 = -\varphi_{14} p_h + \varphi_{24} E_{cool} F_2\, (1 + R_5) Q_5 \\ &= \varphi_{35} R_3 Q_3 \end{aligned} \quad (6)$$

From which we can determine the value of the maximum heating temperature of the product

$$T_2 = \sqrt[4]{\frac{E_{02}}{\sigma_0}} \quad (7),$$

having previously obtained the calculated expression for E_{02}

$$E_{02} = A^* \frac{p_h}{F_2} + E_{cool} \quad (8),$$

Where

$$A^* = \frac{1 - \varphi_{14} + (\varphi_{42} + \varphi_{43})R_4 + \frac{A_{12}^D}{A_{11}^D}(\varphi_{12} + \varphi_{42}R_4)}{\varphi_{24} + \frac{A_{12}^D}{A_{11}^D}} \quad (9),$$

$$A_{12}^D = \varphi_{32}R_3 - \varphi_{42}R_4 - \frac{\varphi_{42}R_4\varphi_{35}R_3}{1 + R_5} \quad (10),$$

$$A_{12}^D = \left(1 + \frac{\varphi_{35}R_3}{1 + R_5}\right)(1 + (\varphi_{42} + \varphi_{43})R_4) + \varphi_{34}R_3 \quad (11).$$

The value of A^* depends on the geometric *(φ_{12}, φ_{13}, φ_{23}, φ_{32}, φ_{35}, φ_{52}, φ_{53}, φ_{54})* and thermophysical *(R_3, R_4, R_5,)* parameters of the system under consideration. To calculate the view factors, we introduce an auxiliary surface F_0, equal to the area of the reflector output section surface. The view factors φ_{10}, φ_{12}, φ_{20} are determined according to the schemes presented in Figure 2 [6, 7]. The remaining coefficients are determined by the algebraic method.

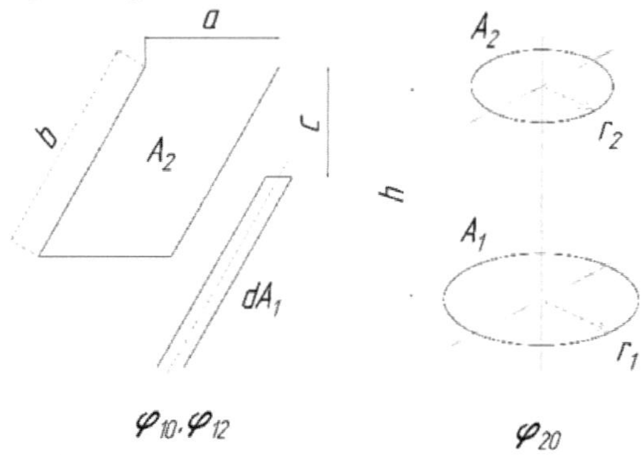

Figure 2. Schemes for determining the base view factors

The input parameters of the model: the distance between the output section of the reflector and the substrate *h*, mm; the area of the substrate (substrate holder) F_{sub}, mm^2; the internal surface area of the reflector (considered equal to the external one) F_{ref}, mm^2; integral emissivity coefficients of the internal surface of the chamber, internal and external surfaces of the reflector ε_3, ε_4 and ε_5, respectively; the cooling liquid temperature T_{cool}; the chamber internal surface area F_{cham}, mm^2; the heater power p_h, W. Graphs of the substrate temperature dependence on some of the input parameters were plotted, while the remaining parameters were recorded in accordance with their values on the existing experimental setup. The following values were taken as the base ones: $h=30$ *mm*, $F_{sub}=16742$ *mm*^{2} (diameter *146 mm*), $F_{ref}=23095$ *mm*^{2}, $\varepsilon_3=\varepsilon_4=\varepsilon_5=0{,}075$ (polished stainless steel), $T_{cool}=25°C$, $F_{cham}=$ *158060 mm*^{2}, $p_h=100$ W. The substrate temperature dependence on *h* is shown in Figure 3.

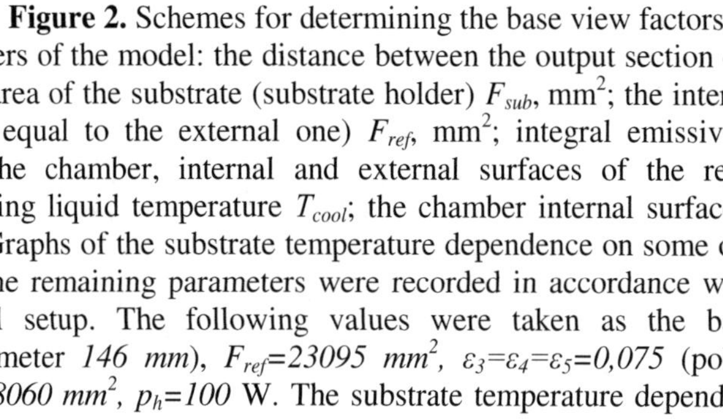

T, °C	369	357	345	334	326	320	316	314	312
h, mm	0	10	20	30	40	50	60	70	80

IOP Conf. Series: Materials Science and Engineering **781** (2020) 012018 doi:10.1088/1757-899X/781/1/012018

Figure 3. Changes in the substrate temperature with increasing the distance between the output section of the reflector and the substrate

The dependence of the temperature on the substrate radius is shown in Figure 4.

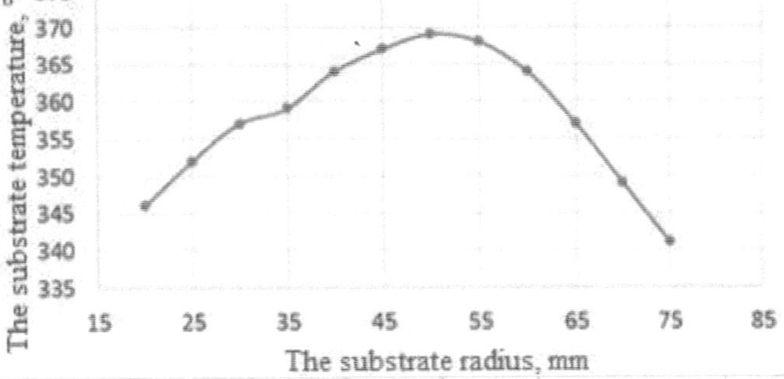

T, ℃	346	352	357	359	364	367	369	368	364	357	349	341
R, mm	20	25	30	35	40	45	50	55	60	65	70	75

Figure 4. Changes in the substrate temperature with increasing the substrate radius

The dependence of the temperature on the emissivity of the chamber walls ε_3 is shown in Figure 5. As can be seen from the graph, the obtained temperature is extremely dependent on the degree of blackness of the chamber, especially in the range from 0,0 to 0,10.

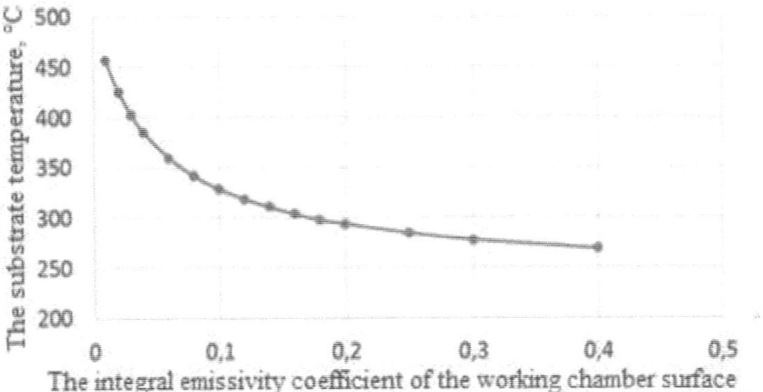

T, ℃	456	425	384	359	341	328	318	303	297	293	284	268
ε3	0,01	0,02	0,04	0,06	0,08	0,1	0,12	0,16	0,18	0,2	0,25	0,4

Figure 5. Changes in the substrate temperature with increasing the integral emissivity coefficient of the working chamber surface

The dependence of the temperature on the emissivity of the working surface of the reflector is shown in Figure 6. It is noteworthy that the reflectivity of the internal surface of the chamber has a greater effect on the substrate temperature than the reflectivity of the internal surface of the reflector.

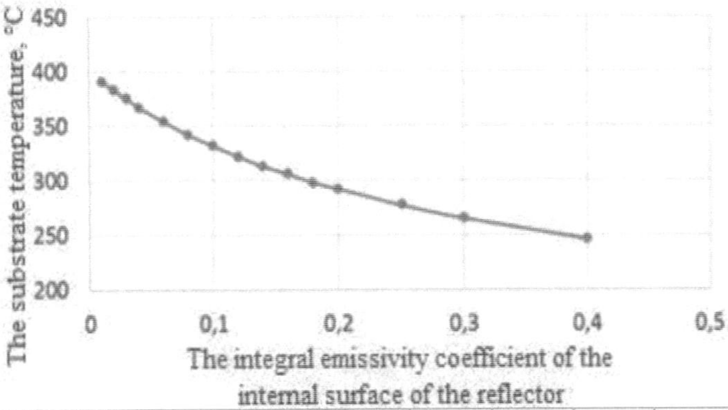

T, ℃	391	383	367	354	342	332	322	306	298	292	277	246
ε4	0,01	0,02	0,04	0,06	0,08	0,1	0,12	0,16	0,18	0,2	0,25	0,4

Figure 6. Changes in the substrate temperature with increasing the integral emissivity coefficient of the internal surface of the reflector

The dependence of the substrate temperature on the cooled walls temperature is shown in Figure 7. As can be seen from the graph, the influence of this factor can be considered small.

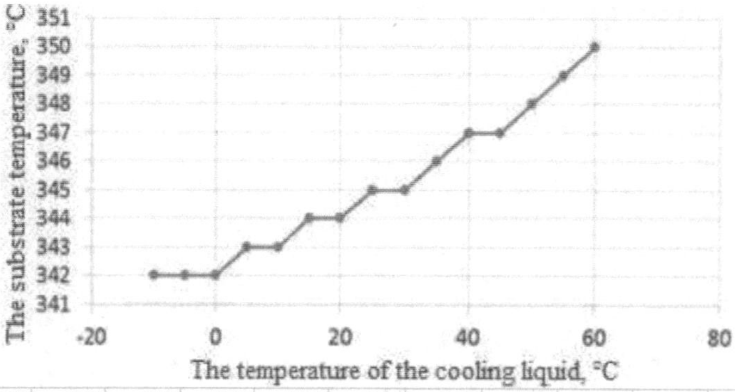

T, ℃	342	342	342	343	343	344	344	345	345	347	348	350
T_{cool}, ℃	-10	-5	0	5	10	15	20	25	30	40	50	60

Figure 7. Changes in the substrate temperature with increasing the temperature of the cooling liquid

The dependence of the temperature on the heater power is shown in Figure 8.

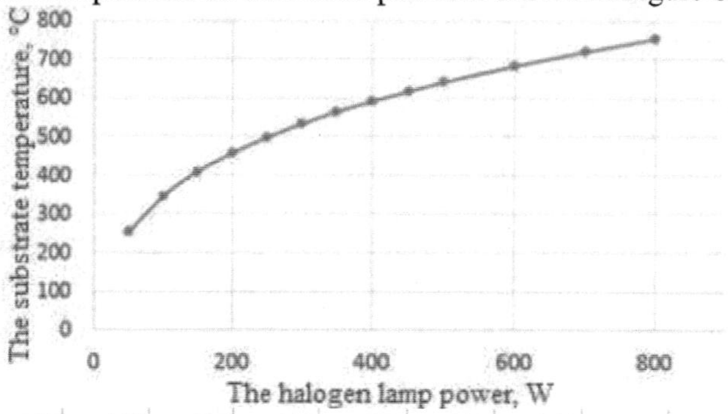

T, ℃	253	345	408	457	498	533	564	592	641	683	720	754
p, W	50	100	150	200	250	300	350	400	500	600	700	800

Figure 8. Changes in the substrate temperature with increasing the heater power

To verify the obtained calculated data, they were experimentally verified on a laboratory equipment. The substrate (substrate holder) with a diameter of 146 mm was heated with three values of the gap between the substrate and the reflector: 30, 40 and 50 mm. The calculated value of the chamber internal surface area was 0,158 m^2. The heating was carried out for 40 minutes. The temperature value was recorded every minute. The graphs of substrate temperature changes over time are shown in Figure 9.

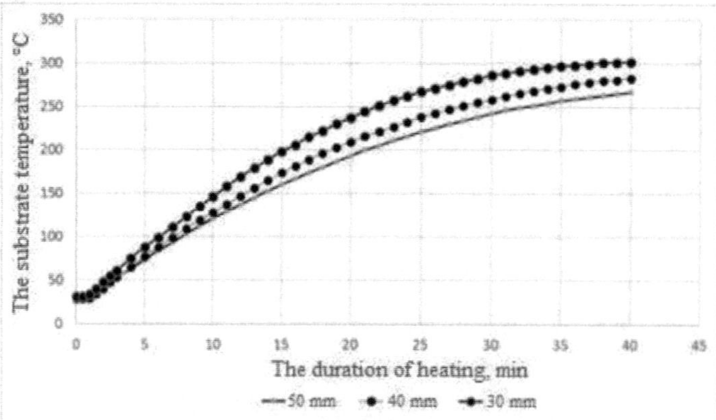

Figure 9. Changes in the substrate temperature over time with three values of the gaps between the reflector and the substrate

Table one compares the theoretical and the experimental data. It may be observed that the difference between the calculated and the experimental values increases with increasing the gap. The reason is that with a 30 mm gap the temperature change practically stopped, which cannot be said about heating with the 40 and 50 mm gaps. It is likely that if the measurements had continued a higher temperature on the substrate for these gap values would have been achieved, the difference of which from the theoretical one would have been similar to the difference with a 30 mm gap in percentage terms.

Table 1. The comparison of the experimental and calculated values of the substrate temperature during heating in vacuum

Gap, mm	30	40	50
Temperature during measurements, ℃	302	283	267
Calculated temperature, ℃	336	328	322
Calculation error, %	11	16	20

The conducted studies allow us to speak of a sufficiently high accuracy of the developed model, taking into account the wide range of published data on the values of the materials emissivity and the lack of capabilities to measure them during the experiments.

References

[1] Sergey Reznik. Thermal regimes of space composite structures. Part I // Heat and Mass Transfer in the Thermal Control System of Technical and Technological Energy, 2018, Vol.194.

[2] D. E. Kablova, V. V. Sidorova, Yu. A. Puchkov. Thermodynamic and Kinetic Aspects of the Phosphorus Effect on the Stability of the Structure and Properties of Single-Crystal Refractory Nickel Alloys // (2019) Inorganic Materials: Applied Research, 2019, Vol. 10, No. 4, pp. 979–986.

[3] Sergey Reznik. Thermal regimes of space composite structures. Part II // Heat and Mass Transfer in the Thermal Control System of Technical and Technological Energy, 2018, Vol.194.

[4] V.V. Krapukhin, V.G. Kosushkin, L.V. Kozhitov, V.G. Kostishin, D.G. Muratov, A.V. Popkova. Evolution of Models and Algorithms for Calculating the Parameters of Technological Processes to Obtain Materials for Micro- and Nanoelectronics // Russian Microelectronics, 2017, December, Vol. 46, Issue 8, pp. 571-579.

[5] Kuvyrkin G N, Zhuravskii AV, Savel'eva I Y. Mathematical Modeling of Chemical Vapor Deposition of Material on a Curvilinear Surface // Journal of Engineering Physics and Thermophysics, 2016, 89 (6), pp. 1374-1379.

[6] R. Siegel, J.R. Howell. Thermal radiation heat transfer // 2001, P. 934.

[7] E.M. Sparrow, R.D. Gess. Radiation Heat Transfer // 1978, P. 296

IOP Conf. Series: Materials Science and Engineering **781** (2020) 012019 doi:10.1088/1757-899X/781/1/012019

Features of Au(80)Sn(20) thin film solder manufacturing technology

E M Filonenko, A V Fomin and V V Ivanov

E.I. Zababakhin All-Russia Research Institute of Technical Physics-Russian Federal Nuclear Center, 456770, Snezhinsk, Russia

E-mail: dep5@vniitf.ru

Abstract. This work provides the results of manufacturing technology development for Au(80)Sn(20) thin film solders produced with electron-beam evaporation coating of alternating gold and tin layers. The phase formation mechanism in the produced solders was studied to work out an evaporation procedure which enables to make a solder with the properties meeting the requirements to soldering of a laser diode onto a heatsink base surface.

1. Introduction

An Au80Sn20 alloy with a melting point of 280 °C is a eutectic in the gold tin system [1]. This alloy is one of the critical materials used to manufacture state-of-the-art optoelectronic devices. In particular, its unique properties can be applied to fabricate semiconductor lasers with high power laser diodes (LD). Foils, tapes, preforms, and soldering pastes represent today's Russian commercial varieties of Au(80)Sn(20) [2] but for flip-chip die soldering, the best option to be used is its thin-film variety deposited directly on the components to be joined

Within the context of the semiconductor lasers production, one of the key issues was to develop an in-house manufacturing technology for an Au(80)Sn(20) solder comprising layers less than 10 μm which are deposited directly on heatsinks. For this purpose, electron-beam deposition of alternating gold and tin layers was employed. This method is intended to make micro-relief films and to control the solder stoichiometric composition to a high accuracy.

2. Electron-beam Evaporation Coating System

For vacuum evaporation of Au(80)Sn(20) solders, as well as buffer and adhesion layers, an electron-beam evaporation system was used. The system incorporates quartz measuring heads to monitor the thicknesses of alternating gold and tin layers to a high accuracy. Fig.1 shows an example of the system vacuum chamber configurations.

The heatsinks are placed and moved within the chamber by means of a rotary segmented calotte. The rotary segmented calotte design is given in Fig.2. It comprises three spherical surface segments, each of which has 8 holes to mount the heatsinks with the substrate holders.

Studies of the obtained solders allow us to state that the selected design concept of the system inner chamber provides uniform coatings having diversity in thicknesses of less than ±5%. Besides, the sequentially loaded heatsinks might amount to 4500 pieces.

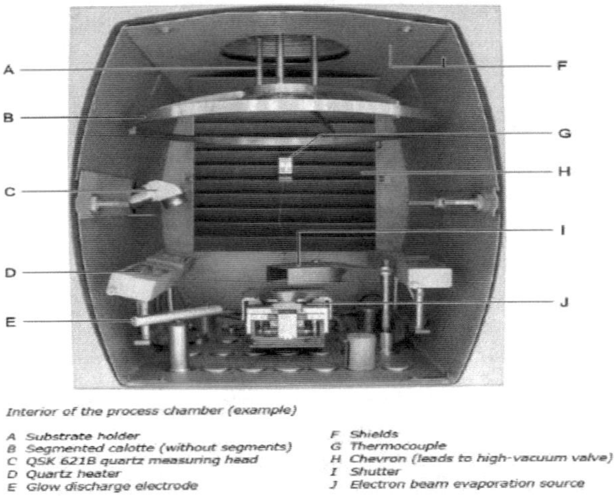

Figure 1. Vacuum Chamber Configuration: A- calotte holder, B- rotary segmented calotte (without segments), C- quartz measuring head, D – quartz heater, E - glow discharge electrode, F - shields, G - thermocouple, H - vacuum system chevron, I - shutter, J - electron beam evaporation source with 6 crucibles.

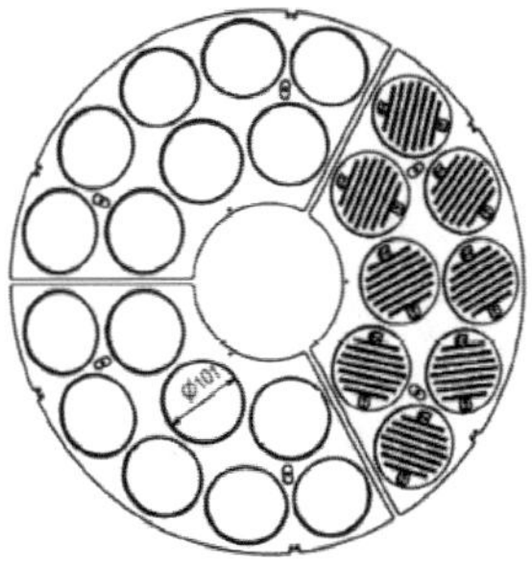

Figure 2. Image of a three-segment rotary calotte with holes and substrate holders.

3. Features of Au(80)Sn(20) Solder Manufacturing

According to [4], the alloy Au(80)Sn(20) is a eutectic, comprising Au_5Sn (ξ), and AuSn (δ) phases at room temperature. The alloy microstructure incorporates ξ and ξ' dendrites - Au_5Sn surrounded by (ξ+δ) composition matrix.

In the described solder manufacturing method used to deposit alternating gold and tin layers, the phases are being formed directly during the soldering procedure while the obtained composite is being heated. Homogeneity of ξ and δ phases' distribution through the solder bulk will depend on the alternating layers' thickness ratio and will control the solder behavior for the full duration of soldering.

Various evaporation procedures, including the existing ones were studied to find the best ratio of gold and tin layers [5]. The obtained solders were analyzed to determine their melting character. In the studies, the temperature range together with the solders' heating rate were selected as relevant to the laser diode soldering with Au(80)Sn(20) solder – (280 – 340) °C.

The findings of the study, including visual, X-ray, and differential scanning calorimetric analyses enabled to formulate unambiguous criteria to select the ratio of the thicknesses for the alternating layers in the solder composition. The solder melting character in case of an unbalanced alternating gold and tin layers' ratio can be exemplified by a 3-μm thick Au(80)Sn(20) solder obtained with one of the studied evaporation procedures.

It was noted that when passing through the eutectic-formation temperature (280 °C), the melt is being formed. It remains unchanged up to 300 °C. The temperature rise above 300°C or the exposure time of more than 5 seconds at 300°C resulted in the emergence of local crystallization areas and further rapid crystallization through the solder bulk. Figure 3 shows typical images of the sample coating surfaces produced after reflowing at 300 °C during 5 seconds. By means of X-ray spectrometry it was determined that the local crystallization areas were large crystals of δ phase. This implies deviations form the eutectic alloy microstructure.

Images of the solder cross-sections exhibit partial layering of the eutectic-forming phases through the solder bulk, and δ phase is predominantly concentrated at the solder-barrier metallization interface (Fig.4).

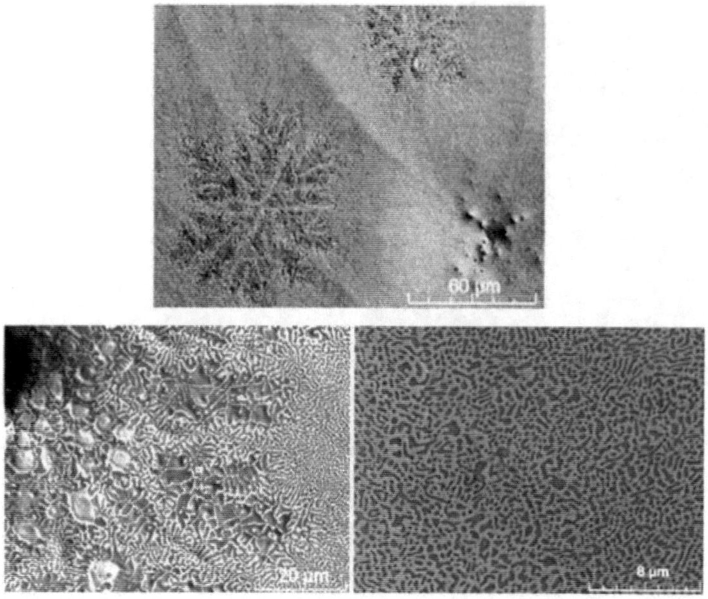

Figure 3. The back-scattered electron images of a 3-µm thick AuSn solder exposed at 300°C during 5 seconds show the following areas a) local crystallization areas, b) eutectic structure.

The observed inhomogeneous distribution of the phases through the eutectic structure and the visual analysis of the reflowing stages imply that the obtained solder is not homogeneously melted at the eutectic-formation temperature or above.

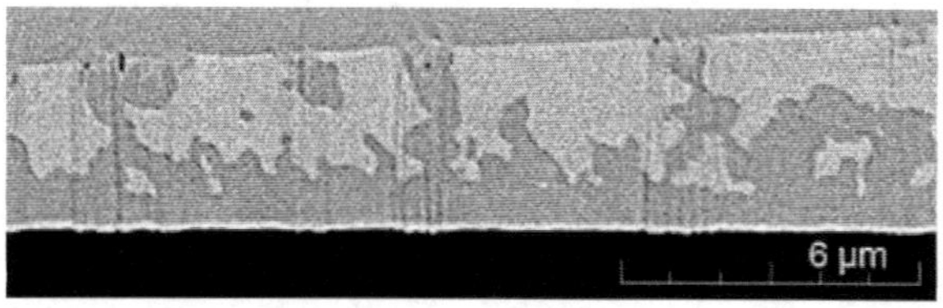

Figure 4. The electron back-scattered image of a 3-µm thick Au(80)Sn(20) solder exhibits inhomogeneous distribution of the eutectic-forming phases through the solder bulk.

Similar solders were used to mount the laser bars on the heatsinks. However, because of the peculiarities quoted above the laser bar metallization was only partially wet; soldering procedure, in terms of its reproducible quality, was inadequate (Fig.5).

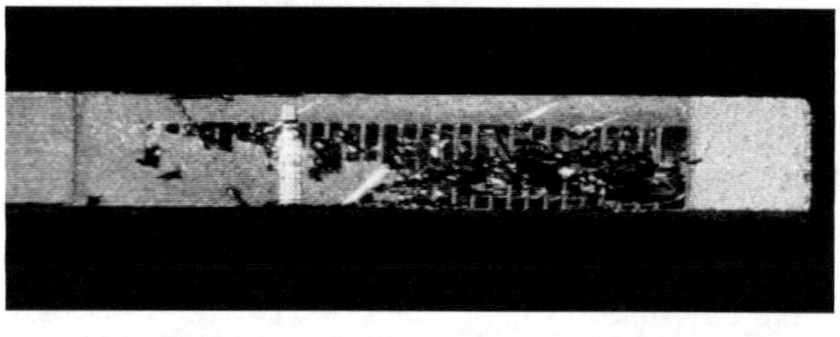

Figure 5. Visual appearance of a solder joint with the laser bar removed. The area to the left shows that the laser bar metallization is not wet with the Au(80)Sn(20) solder, the stripes are traces of the laser bar emitters' contact, the arbitrary shaped dark and light areas are the laser bar fragments which have remained after detachment.

The method was further developed to improve the melt stability at the soldering process due to modification of the existing evaporation procedures used for gold and tin alternating alloys. The best ratio for the solder alternating components' thicknesses was found, and the required evaporation temperature conditions for the layers deposition [7] were identified considering the recorded data on the diffusion processes in the couple gold and tin [6]. As for the developed procedure, the thickness of the obtained Au(80)Sn(20) solder made (5.5-6) μm.

The solders were analyzed by differential scanning calorimetry to demonstrate that they start to melt at the eutectic-formation temperature (280°C) and then the melt remains steady up to 340°C (Fig. 6a). The electron microscopy was used to study the surface and cross-section of the solders after reflowing at 340°C. The analysis revealed a typical eutectic structure of the alloy AuSn at room temperature (Fig. 6b).

The techniques used to mount a laser bar on the heatsinks were worked out with this thin film solder. Studies of the obtained solder joints between the laser bar and the heatsinks convey the following: the laser bar metallization is well wet by the obtained eutectic solder and the eutectic-forming phases are homogeneously distributed in the solder (Fig. 7).

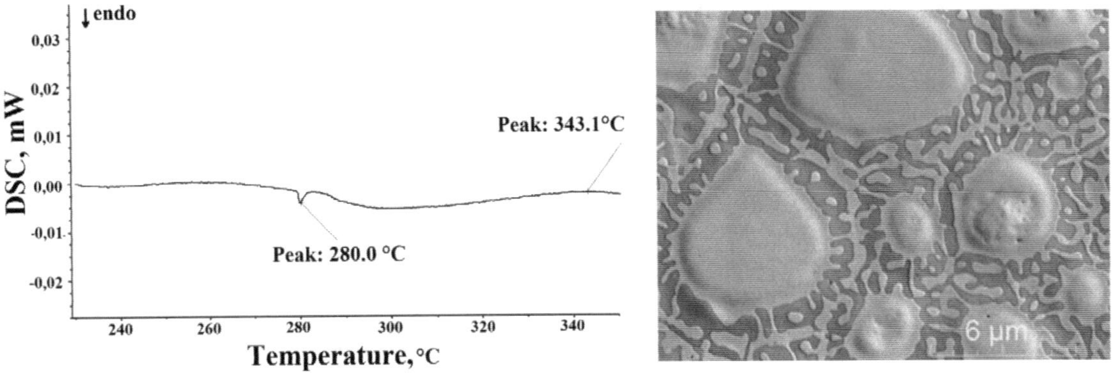

Figure 6. Results of analysis for 5.5-6 μm-thick Au(80)Sn(20) solders: a) – DSC-curve recorded in the temperature range (100–400) °C; b) – a typical image of a 5.5 μm thick Au(80)Sn(20) at 340 °C demonstrates a eutectic structure: ξ phase dendrites surrounded with (ξ+ δ) matrix.

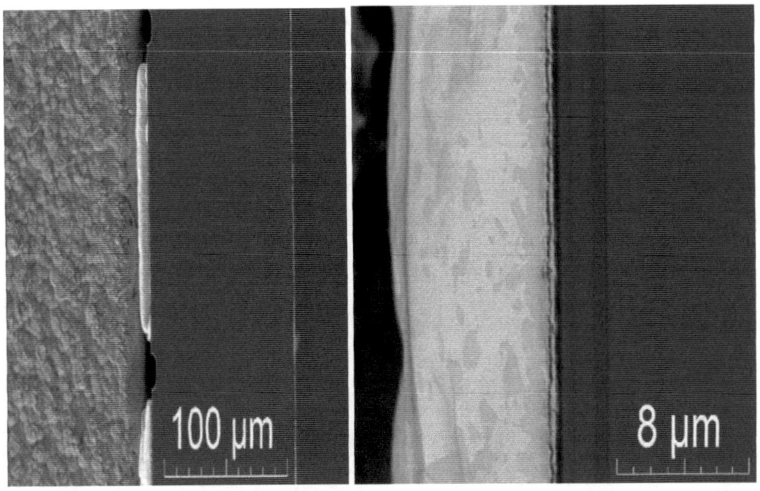

Figure 7. Image of a solder joint between a laser bar and a heatsink with the surfaces (to be joined) properly wet by the solder.

4. Conclusions

Electron-beam evaporation coating for alternating gold and tin layers was successfully used to develop an Au(80)Sn(20) thin film solder manufacturing method. The selected design concept of the system internal chamber provided uniform deposition of solders on the heatsinks of lasers. Diversity in the thicknesses of the deposited coatings was less than 5%.

The solder evaporation procedure was developed through the studies of Au(80)Sn(20) solders deposited using various ratios of the thicknesses of alternating gold and tin layers. This procedure gave us a Au(80)Sn(20) solder with the total thickness of 5.5 μm and the composition having the eutectic structure within the gold-tin system. At the same time, distribution of the eutectic-forming phases through the bulk of the obtained solder is homogeneous for the full duration of the laser bar soldering.

The developed technology was successfully implemented to manufacture the semiconductor lasers with high power laser diodes.

References

[1] Lyakishev N P 1997 *State Diagrams of Binary Metal* (Moscow: Mashinostroenie) 400
[2] Kondratyuk R 2015 *Electronics: science, technology, and business* **10** 154–160
[3] McNulty J.C. 2008 *Int. Microelectronics and Packaging Society* 909-916
[4] Matijasevic G S, Lee C C., Wang C Y 1993 *Thins solid films* **223** 276-287
[5] Katz A. et al. 1993 *Pat. 5197654 USA*
[6] Yamada T. et al. 2004 *J. of Material Science* **39** 2327 – 2334
[7] Fomin A V, Smirnov E V, Filonenko E M 2019 *Pat. 2691152 Russia*

Investigation of MoS_2 ultrathin films formed by physical vapor deposition in vacuum

A I Belikov, Kyaw Zin Phyo, A I Syomochkin

Department of Electronic Technologies in Mechanical Engineering, Bauman Moscow State Technical University (BMSTU), 105005 Moscow, Russia

E-mail: belikov@bmstu.ru

Abstract. The article presents the optical spectroscopy results of molybdenum disulfide ultrathin films deposited on Si substrates by the methods of direct current (DCMS) and pulsed (PMS) magnetron sputtering MoS_2 target in vacuum, and also at flow deposition under the influence of external magnetic field on the substrate. Furthermore, the article revues the results of optical bandgap calculation for all samples of MoS_2 thin films.

1. Introduction

The investigation of the two-dimensional (2D) materials is one of the budding areas of nanoelectronics for creating experimental devices with high functional properties and minimal energy consumption. In recent years researches on 2D-material such as graphene were phenomenally successful [1]. Due to its unique electronic properties, graphene attracts the interest of researchers as a 2D material for nanoelectronic's area [2]. The disadvantage of graphene is the absence of the bandgap energy in its electronic structure. This fact is an obstacle for a number of applications in the field of optoelectronics and discrete electronic. This limitation has led to the search and investigation of an alternative 2D materials that exhibit semiconductor properties. Such materials typically have a crystalline structure similar to graphite. The presence of weak "Van-der-Waals" forces between atoms located in different layers and strong bonds within the layers, allowed the researchers to separate weakly bound layers and obtain the first graphene samples. Molybdenum disulfide (MoS_2) belongs to the group of transition metal dichalcogenides (TMDs). Molybdenum disulfide (MoS_2) has a resembling graphite layered structure. MoS_2 is endowed with semiconducting properties and this material has been the subject of intensive research in recent years.

Molybdenum disulfide is one of the most studied farsighted materials for nanoelectronic and optoelectronic devices due to its unique electronic, mechanical, and optical properties [3]. The bulk crystal of molybdenum disulfide is a semiconductor with an indirect bandgap of 1.24 eV. With a decrease in the number of MoS_2 monolayers, the bandgap width increases. The bandgap width value for molybdenum disulfide monolayer reaches the value of $\approx$ 1.8 eV. An important characteristic of semiconductors is a charge carrier mobility, which determines the frequency characteristics of electronic devices [4]. The high charge carrier mobility in the molybdenum disulfide layers is according to the zero mass of the ambipolar electron-hole pairs. The maximum current carrier mobility of a MoS_2 monolayer reaches 500 $cm^2 \cdot V^{-1} \cdot s^{-1}$. The transparency of the molybdenum disulfide monolayer is characterized by a high optical transmittance of $\approx$ 98.3%, which exceeds the known value for graphene (97.7%) [6]. This set of properties determines the promising possibilities of using molybdenum disulfide with graphene as element base materials for thin transparent semiconductor systems.

The relevant objective is to develop methods that provide high-quality, homogeneous, textured MoS_2 ultra-thin films on large-sized substrates. Magnetron sputtering of a cathode target is one of the most future methods of applying MoS_2 thin films. The physical vapor deposition (PVD) method, and especially magnetron sputtering, promotes the growth of films with the desired grain size, thickness and morphology. Bulk molybdenum disulfide is a non-magnetic material. However, the molybdenum disulfide molecule is asymmetric, and each electron spin has its own magnetic moment. It is assumed that the intrinsic magnetic moment of MoS_2 molecules can be used to realize orientation effects during film deposition under the conditions of an applied magnetic field. The influence of an external magnetic field will orient the molecules of MoS_2 and integrate them, into the structure, forming the texture of the film. The aim of this work was to assess the influence of an external magnetic field, applied in the region of the silicon substrate surface during magnetron sputtering on the optical and electronic properties of ultrathin MoS_2 films.

To determine the optical characteristics of thin films, various methods such as Raman spectroscopy, spectroscopic ellipsometry, optical spectroscopy, and others are used. We used the method of the optical spectroscopy of thin films on the surface of a semiconductor material substrate. This method is based on measuring the reflection coefficient spectral values for the wavelength range. Optical spectroscopy is also the most common way to determine the optical bandgap.

2.Research methods

To calculate the bandgap width of MoS_2 thin films we used a methodology that includes four main steps.

At the first stage of this methodology we study the reflectivity of the MoS_2 thin film deposited on the Si-substrates. The measurement results are presented by reflection spectra. In this article, reflection spectra in the wavelength range from 380 to 1100 nm were obtained by using an Isovac Epsilon optical spectrophotometer.

The second step is the calculation of the MoS_2 thin film's absorption coefficients based on the obtained total reflection coefficient (R) of the substrate and thin film sample. The absorption coefficient can be calculated by using the Kubelka-Munk equation (1) [7].

$$F(R_\infty) = \frac{(1-R_\infty)^2}{2R_\infty} = \frac{K}{S} = \frac{2.303\varepsilon C}{S} \quad (1)$$

where K – is the absorption coefficient, S – scattering coefficient

At the third stage, the calculated values of the absorption coefficient are used to construct the dependence of the modified Kubelka-Munk function for semiconductors with an indirect bandgap $(F(R_\infty) \times h\vartheta)^{1/2}$ on the photon energy hv, in the bandgap energy range from 1.3 eV to 2.7 eV. At the final, fourth stage, the optical bandgap is determined by the Tauc method. The method is based on a linear approximation of the modified semiconductor absorption coefficient dependence on the probe radiation energy near the low-frequency edge of the fundamental absorption band.

3. Experiments and results

To prepare samples with MoS_2 films, we used an experimental setup that ensured the methods implementation of constant (DCMS) and pulsed (PMS) magnetron deposition. The installation is equipped with balanced magnetron sputtering systems (MSS) with power sources that provide stabilization of magnetron operating modes. The diameter of the cathode target was 75 mm and a thickness of the sputtering target was 5 mm. In our experiment, the target purity was 99.9%. Fragments of semiconductor silicon wafers with a crystallographic orientation of [100], 10×10 mm^2 in size, were prepared as substrates. The substrates were preliminary degreased and dried. In the experiments, the following magnetron sputtering factors were varied: presence/absence of the external magnetic field, the magnetron sputtering target mode (direct current or low-frequency pulsed), the substrate temperature and the discharge power, the other parameters remained unchanged. Magnetron sputtering was carried out in argon at a chamber pressure of $5 \cdot 10^{-1}$ Pa. Pulsed magnetron sputtering was carried out at a discharge power of 28 W, at a frequency of 50 kHz with a 40% pulse duty ratio. The main parameters of the magnetron deposition process are shown in table 1. As a result of the work, 4 test samples of molybdenum disulfide films were obtained.

A magnetic field was applied parallel to the substrates surface. Ferrite magnets were located on the substrate holder by using a specific tooling. Magnets were arranged on the opposite sides of the silicon substrate. The magnetic induction magnitude of the longitudinal magnetic field in the center of the substrate was measured by using a teslameter and amounted to 0.30 T.

Table 1. MoS_2 films magnetron deposition parameters.

№	MSS mode	Discharge power, W	Discharge current, A	Deposition time, seconds	Substrate temperatur e, °C	Cathode-substrate distance, mm	External magnetic field
1	PMS	28	0.05	10	No heating	110	+
2	PMS	28	0.05	10	200	110	–
3	DCMS	22	0.05	10	200	110	+
4	DCMS	22	0.05	10	No heating	110	–

Figure 1 shows the reflection spectra of MoS_2 samples obtained by magnetron deposition in pulsed mode and in direct current mode. Samples obtained under the conditions of an applied external magnetic field showed a higher reflectivity in the entire wavelength range. The reflection spectra nature of these samples is significantly different. The reflectivity of the samples obtained by pulsed magnetron sputtering is lower in comparison with samples obtained at constant current mode.

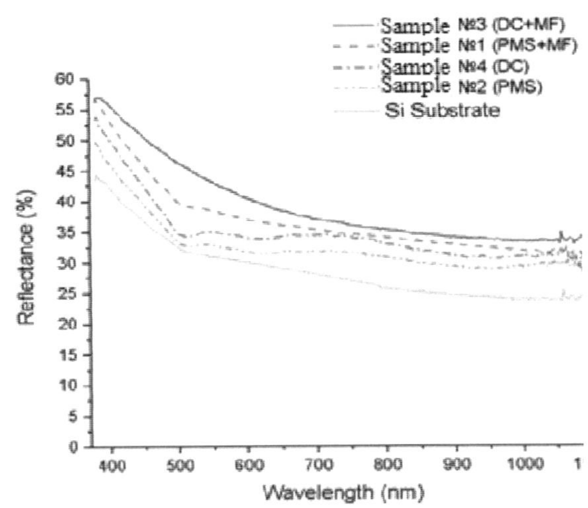

Figure 1. Reflection spectra of MoS_2 thin films.

The absorption spectra (Figure 2) of experimental samples were obtained by using the Kubelka-Munk function (1) that converts the reflection coefficient into absorption coefficient. MoS_2 film samples that were formed without external magnetic fields exhibit more pronounced absorption peaks.

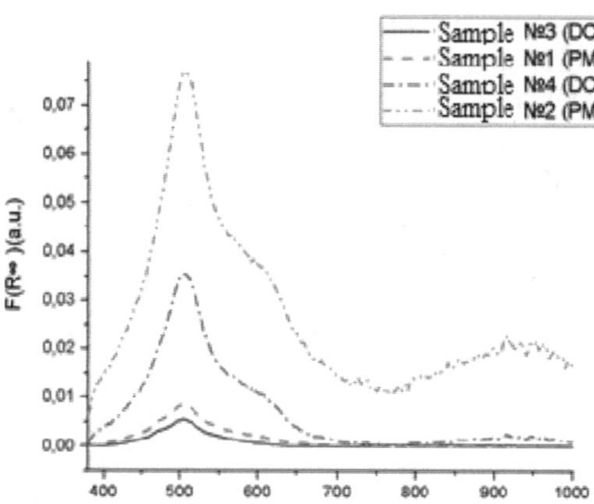

Figure 2. Absorption spectra of MoS_2 thin films.

4. Analysis and discussion

Based on the absorption spectra for all samples, we obtained graphs of the modified absorption function, which are shown in figure 3. In the case of applying an external magnetic field in direct current mode, the bandgap of the MoS_2 film increases from 1.73 eV to 1.78 eV. Similar trend is observed for the pulsed sputtering mode, the bandgap increases from 1.71 eV (sample obtained without external magnetic field) to 1.76 eV (sample obtained with external magnetic field). The modified absorption functions of the samples deposited under the influence of an external magnetic field have a pronounced absorption edge, which may indicate a higher degree of perfection and texture of the MoS_2 thin films.

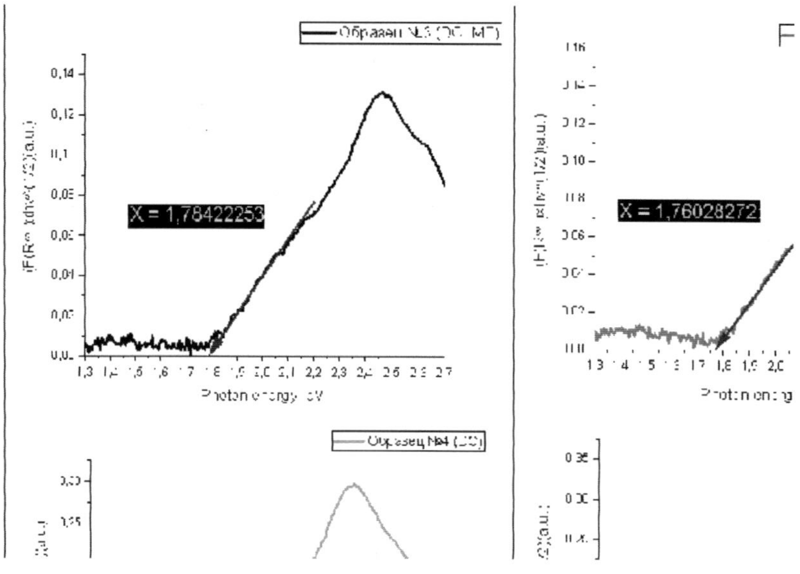

Figure 3. Tauc plots method for the bandgap energy determination of MoS_2 thin films experimental samples.

With the use of the atomic force microscopy (AFM Solver-NEXT), was carried out surface morphology studies and evaluated thickness of MoS_2 thin films on prepared experimental samples. A summary of the bandgap values and the samples thickness are presented in table 2.

Table 2. MoS_2 thin films samples parameters.

№	MSS mode	External magnetic field	Bandgap width, eV	Thickness, nm
1	PMS	+	1.76	5.6
2	PMS	–	1.71	4.2
3	DCMS	+	1.78	4.1
4	DCMS	–	1.73	3.5

5. Conclusions

As a result of our investigation, a significant effect of a longitudinal external magnetic field applied to the surface of a silicon substrate during magnetron deposition on the electronic and optical properties of ultrathin MoS_2 films was found. The obtained modified Kubelka-Munk absorption functions indicate the presence of a clear absorption boundary for MoS_2 film samples, which were deposited under the influence of an external magnetic field, which probably indicates their higher structural perfection. The same deposition time leads to a significant increase in the film thickness (by 20-30%) for cases of the presence of a longitudinal magnetic field with induction of about 0.3 T at the substrate surface during deposition, this fact indicates about the applied external magnetic field influence on the rate of growth films.

References

[1] Novoselov K S, Geim A K, Morozov S V, Jiang D, Zhang Y, Dubonos S V, Grigorieva I V, Firsov A A 2004 *Science***666** 306
[2] Papageorgiou D G, Kinloch I A, Young R J 2017 *Prog. Mater. Sci.* **90** 75
[3] Duan X et al. 2015 *Chem. Soc. Rev* **44** 8859
[4] Lopez Sanchez O et al. 2013 *Nat. Nanotechol***8** 497–501
[5] Baugher B W et al. 2013 *Nano letters***13** 4212-4216
[6] Lee P A: Optical and *Electrical Properties***4** 1976 464
[7] Arney J S, Chauvin J, Nauman J, Anderson P G J. 2003 *Imagin Sci. Technol***47** 339-345

IOP Conf. Series: Materials Science and Engineering **781** (2020) 012021 doi:10.1088/1757-899X/781/1/012021

Ultrathin film deposition for nanoelectronic device manucturing

Y V Panfilov, I A Rodionov, I A Ryzhikov, A S Baburin, D O Moskalev and E S Lotkov

Moscow, Bauman Moscow State Technical University, 5, 2-d Baumanskay St., MT-11 Email: panfilov@bmstu.ru

Abstract Study results of thin film with thickness less than 50 nm deposition by means of electron beam evaporation on special vacuum coaters were represented. Calculation results of energy and mass carry magnitude for ITO, Ag and Al thin films deposition with preset of structure and composition homogeneity and roughness were shown.

1. Introduction

Thin film deposition opportunity with thickness less than 50 nm and preset of quality characteristics depends on energy and mass carry magnitude [1]. Thin film generation particles energy depends on thin film deposition method, and thin film generation particles mass flow depends on source conditions and distance from the source and substrate. So, ultrathin and ultrasmooth and uniform thin films with crystal structure and homogeneous composition can be deposited by means of a little deposition rate and particle energy, substrate high temperature and high vacuum as well.

Electron beam evaporation method corresponds to all of the enumerated requests [2]. The science and education center of Bauman Moscow State Technical University has the special vacuum electron beam equipment for nanostructured thin films manufacturing of nanoplasmonic and quantum calculation devices for modern nanoelectronics.

2. The special vacuum technological equipment

The vacuum coater «EvoVac Angstrom Engineering» consists of pumpdown system (spiral forevacuum and cryogen high vacuum pumps) and technological sources (electron tween beams evaporator with seven melting pots and ion source with hollow cathode) and vacuum chamber with height adjusting and turning with forty revolutions per minute 200-mm substrate holder. The substrate holder can be heated up to 773 K and can be cooled up to liquid nitrogen temperature as well. Also there are two sensors of thin film deposition rate and film thickness optic monitoring into the vacuum chamber. Unevenness of thin film thickness on 200-mm substrate unexceeds of ± 2 %.

The vacuum electron beam evaporating coater Plassys MEB550 SL3 is intended for metallic and oxide thin films deposition. The preset of thin oxide film stoichiometry is provided by means of oxygen flow supply into vacuum chamber control. Substrates can be heated up to 873 K and substrate surface can be cleaned before thin film deposition by means of argon plasma discharge. The coater consists of three chambers with the next pressure in it: $3 \cdot 10^{-6}$ Pa in loading chamber, 10^{-7} Pa in metallic thin film deposition chamber, 10^{-6} Pa in oxidation chamber. All of three chambers are pumped by means of Edwards spiral forevacuum and turbomolecular high vacuum pumps.

There are 100-mm substrate holder with three variance and automatic control system for thin film deposition on preset algorithm at the coater.

IOP Conf. Series: Materials Science and Engineering **781** (2020) 012021 doi:10.1088/1757-899X/781/1/012021

3.Electron beam ultrathin films deposition examples

10 – 20-nm Indium Tin Oxide (ITO) thin films are used at optic modulators as a transparent conducting oxide. Method of ITO thin film deposition was electron beam evaporation. In_2O_3 and SnO granules were evaporated by electron beam with a pressure of 10^{-5} Pa and were deposited on silicon substrate. There was ion assistance of thin film condensation and 0,2 nm/sec thin film deposition rate was controlled by means of a quartz sensor.

The main element of superconductive cubits for quantum calculation is josephson's junction (Fig. 1) – three layers structure as "superconductor – dielectric – superconductor".

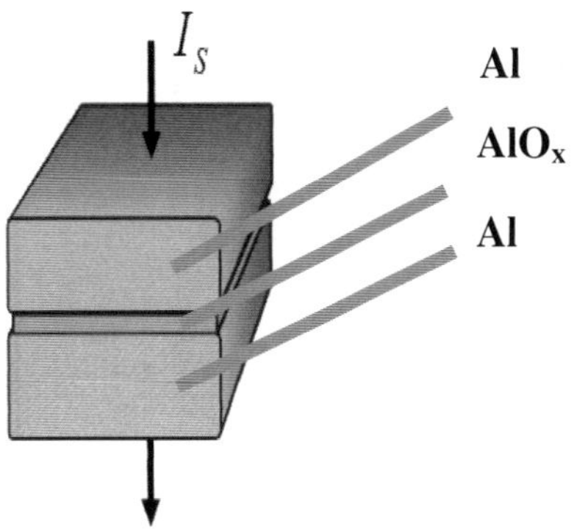

Figure 1. Jjosephson's junction scheme (I_s electric current)

100-nm thin aluminum film as a superconductor was deposited by electron beam evaporation method on Plassys MEB550 SL3 coater. Unified vacuum cycle of thin film deposition with 2 – 10 Å/sec rate was used for providing precise manufacture of 1,5-nm dielectric layer as a tunnel barrier.

Manufactured thin films had uniform of 100% and roughness less than 2 nm. Thin film deposition rate influence to thin film roughness is shown at Fig. 2. Received experimental results point out that thin film roughness minimum can be deposited on 0,5 nm/sec rate.

40-nm silver films with mono crystal structure and uniform of 100% and roughness less than 1 nm are the best material for nanoplasmonics. Such films was manufactured by means of EvoVac Angstrom Engineering coater on 10^{-5} Pa pressure and substrates temperature of 400 – 800 K and thin film deposition rate of 0,1 – 0,4 nm/sec as well.

It is possible to provide thin film deposition rates of 0,1 – 0,4 nm/sec by electron beam evaporation only if material temperature in melting pot corresponds with preset saturated vapor pressure of thin film material.

IOP Conf. Series: Materials Science and Engineering **781** (2020) 012021 doi:10.1088/1757-899X/781/1/012021

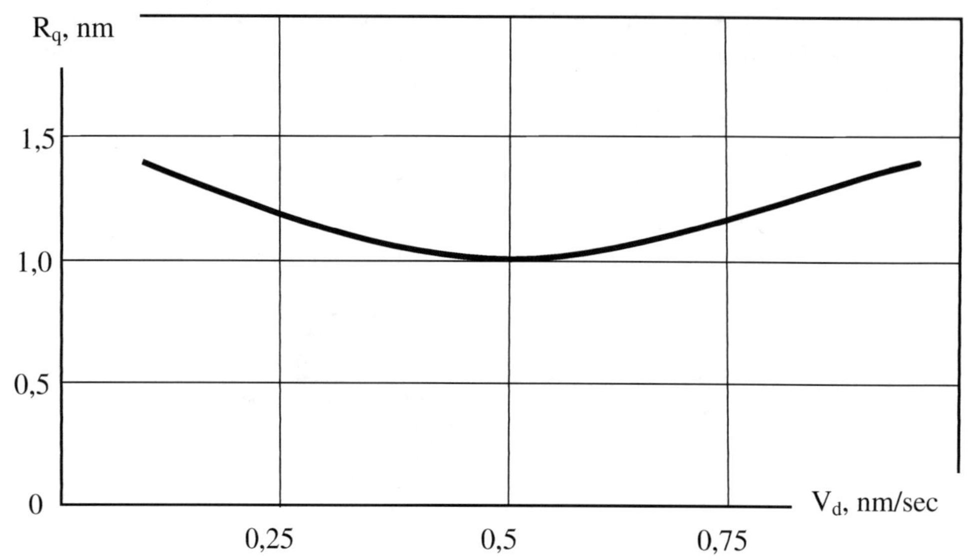

Figure 2. The influence of thin film rate V_d to thin film surface roughness R_q

The experimental research of an influence of silver crystal dimension from substrate temperature (Fig.3) and thin silver film rate (Fig.4) were carried out. The theoretical dependence of aluminum and silver thin film deposition rate V_d from heating temperature of thin film materials up to evaporation temperature T_e with corresponding saturated vapor pressure are represented on Fig. 5.

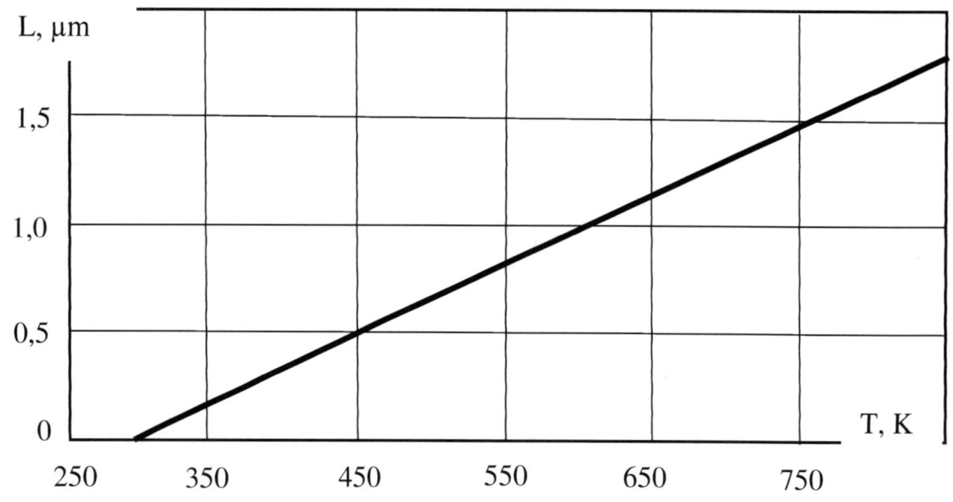

Figure 3. The influence of silver crystal dimension L from substrate temperature T

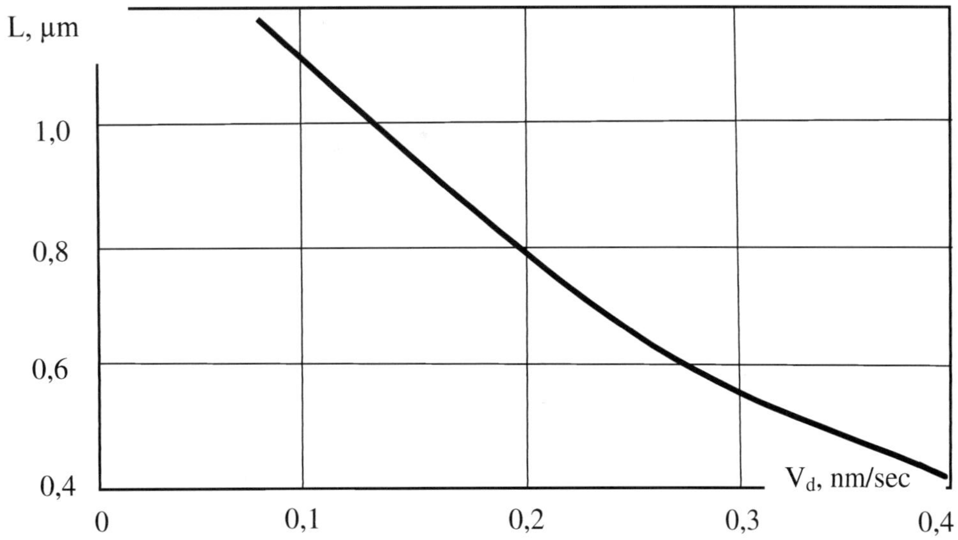

Figure 4. The influence of silver crystal dimension L from thin film deposition rate V_d

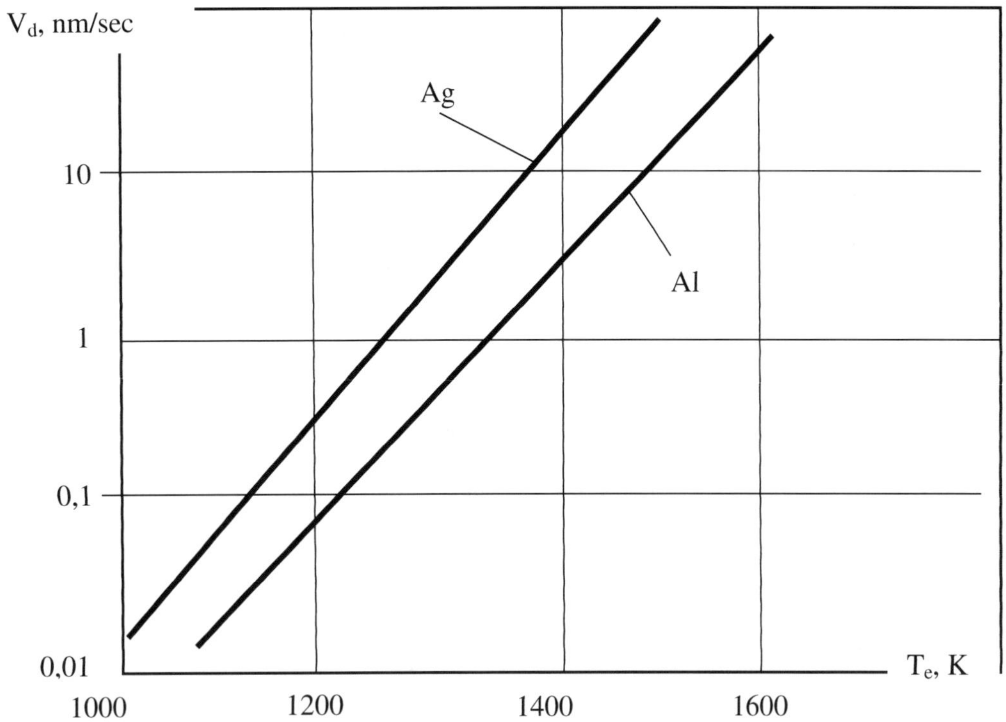

Figure 5. The influence of thin aluminum and silver films rate V_d from evaporation temperature T_e

Thin aluminum and silver films deposition rate V_d calculation were shown that to provide the thin films deposition rate of 0,1 – 0,4 nm/sec the electron beam evaporator melting pot temperature T_e must provide 10^{-2} – 10^{-3} Pa saturated vapor pressure. So, thin silver film deposition rate can be approximately equaled

$$V_d = 5{,}83 \cdot 10^{-3} p_s \sqrt{\frac{M}{T_e}\frac{10^9}{\rho}} = 5{,}83 \cdot 10^{-3} 10^{-3} \sqrt{\frac{108}{1195}\frac{10^9}{10500}} = 0{,}16\ {}^{nm}/_{sec},$$

where M is molecule mass, kg/kmol; ρ is material density, kg/m^3.

For thin aluminum film:

$$V_d = 5{,}83 \cdot 10^{-3} 10^{-3} \sqrt{\frac{27}{1355}\frac{10^9}{2700}} = 0{,}3\ {}^{nm}/_{sec}.$$

Energy and mass carry magnitude of thin silver film is equal:

$$E_e V_d = kT_d \cdot 5{,}83 \cdot 10^{-3} p_s \sqrt{\frac{M}{T_e}} = 1{,}38 \cdot 10^{-23} 1195 \cdot 5{,}83 \cdot 10^{-3} 10^{-3} \sqrt{\frac{108}{1195}} = 2{,}88 \cdot 10^{-26} \frac{J \cdot kg}{m^2 sec}$$

or $1{,}8 \cdot 10^{-7}$ eV·kg/(m^{2}sec),

where k is Boltzman constant.

Energy and mass carry magnitude of thin aluminum film is equal:

$$E_e V_d = 1{,}38 \cdot 10^{-23} 1355 \cdot 5{,}83 \cdot 10^{-3} 10^{-3} \sqrt{\frac{27}{1355}} = 1{,}5 \cdot 10^{-26} \frac{J \cdot kg}{m^2 sec}$$

or $9 \cdot 10^{-8}$ eV·kg/(m^{2}sec).

Conclusion

Thin ITO films were deposited by means of electron beam evaporation method on special vacuum coater for transparent electrode with 10 – 20 nm thickness manufacturing. Al-AlO$_x$-Al josephson's junction and thin silver films for nanoplasmonics were manufactured by means of electron beam evaporation method as well. All of 100-nm thin aluminum film and 1,5-nm thin dielectric film and 40-nm thin silver film were deposited with 0,1 – 10 nm/sec deposition rate. All of this thin film had the structure, uniform composition, roughness and thickness unevenness which were corresponded preset characteristics.

So, the results are asserted that to obtain the energy and mass carry magnitude about $1 \cdot 10^{-7}$ eV·kg/(m^{2}sec) for mono crystal thin films manufacture [3] one can by means of electron beam evaporation method. Meanwhile it is necessary to have precise monitoring of thin film deposition rate or to have an opportunity of material at melting pot measurement to provide preset saturated vapor pressure.

References

[1] Y V Panfilov Nanostructured thin film deposition methods selection by “energy mass transfer” criterion / Nanoengineering, 2016, №12, P. 3 – 10.
[2] Y V Panfilov New chapters on thin film deposition methods classification / Handbook. An Engineering Journal, 2016, №5, P. 6 – 15
[3] Monocrystalline films. Under edition of Pinsker, Mir, 1966, 400 pp.

IOP Conf. Series: Materials Science and Engineering **781** (2020) 012022 doi:10.1088/1757-899X/781/1/012022

Development of microfluidic chip manufacturing technology for human sweat analysis

E A Sevryugina[1] and S V Sidorova[1]

[1] Bauman Moscow State Technical University, 105005, Moscow, Russia

E-mail: katesevr@gmail.com, sidorova@bmstu.ru

Abstract. A technology for manufacturing a microfluidic chip for analyzing the composition of human sweat is proposed. Microfluidic chips can be made of various materials, and, depending on this, can be made by various methods. The analysis of the materials of the constituent components of the product. Equipment was selected for each operation, modes were selected.

1. Introduction

A microfumed chip (also called a lab-on-chip) is a miniature device designed for instant analysis of biological fluids. One of the advantages of this technology is the small volume of liquid for analysis, which makes it very easy to use in medicine and veterinary diagnostics [1].

The project is supposed to modernize the existing analogue, to develop a technology for microfluidic chip production. Unlike the chip for blood analysis, the new development is designed for the analysis of sweat. Changes to the object of analysis will avoid some of the shortcomings of the existing technology.

One of the important parts of the device being developed is the conductive layer. This layer is necessary for transmitting a signal from a biomarker to an electronic device, and is obtained by applying a thin film on the base material.

This chip is supposed to be manufactured using vacuum technological methods. This will increase the purity of the internal surface of the microchannels. Also, a vacuum is necessary for applying a conductive layer, which is responsible for transmitting the change in resistance, when an analyte is in contact with it, on an analytical device. This will ensure the absence of contamination of the material and allow you to apply a thin layer of metal with the necessary range of resistance.

2. Structure of the microfluidic chip under development

The microfluidic chip can be divided into 4 main components, shown in Figure 1:

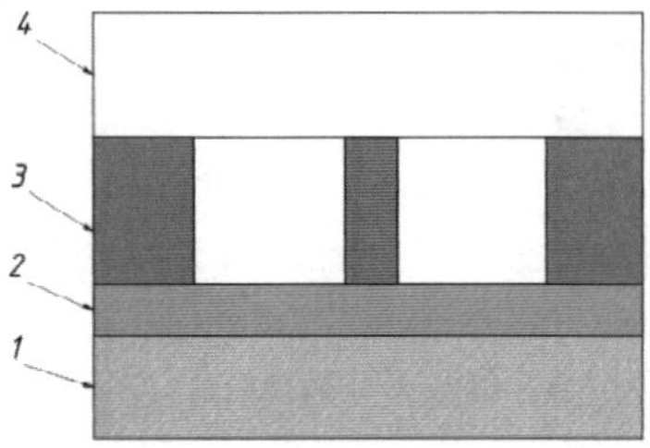

Figure 1. MFC in section. 1- Foundation; 2- Metallised conductive layer; 3- Module; 4- Cover.

The main structure, for analysis, is located in module 3, so the most stringent requirements are imposed on the body material. The main analysis device in a microfluidic chip is a biomarker, which is placed on the conductive layer to read the results from the microfluidic chip to an electronic carrier.

3. Selection of materials components microfluidic chip

Microfluidic chips are made mainly from polymers, less often from glass and silicon, sometimes from metals and ceramics. Polymethyl methacrylate, cycloolefin copolymer, polymethylsiloxane, polyether ether ketone is among the most commonly used polymers. In addition, the module material must be a hydrophobic dielectric that is not susceptible to corrosion. The cover material should also be a hydrophobic dielectric, but its main characteristic is transparency. The cover material should be a transparent plate so that you can observe the movement of the sample under study, as well as monitor the state of the internal structure and replace it in case of a malfunction [2].

The material significantly affects the parameters of the modules. Comparison of materials is presented in Table 1.

Table 1. Analysis of housing materials for the MFC

Parameter	**Polymers**	**Glass**	**Silicon**
		Parameter rating	
Optical properties	Good	Excellent	Good
Mechanical properties	Satisfactorily	Good	Excellent
Chemical resistance	Unsatisfactory	Excellent	Good
Temperature properties	Unsatisfactory	Excellent	Good
Cost effectiveness	Unsatisfactory for prototyping, excellent for large-scale production	Excellent for prototyping, satisfactory for large-scale production	Excellent for prototyping, satisfactory for large-scale production
Recycling	Disposable	Excellent	Satisfactorily
Life time	Satisfactorily	Good	Good

The material significantly affects the parameters of the modules. In laboratory conditions, it is better and more economical to use glass. However, if, in the future, it is planned to release this sample to the market, all research should be carried out immediately on the polymer [3].

Silicon was chosen for the base material, since its parameters (mechanical and temperature properties) satisfy all the conditions set. This material is also suitable for economic parameters.

It is recommended to use polymer for the module material, since it is difficult to find a sample of silicon of the thickness required to obtain a system of capillaries on the material. A polymer can also be used to make the lid, but glass may also be suitable for this role.

The material of the conductive layer must have the highest conductivity, since the film thickness must be the smallest, which is caused by the need to reduce the roughness, as well as the size of the chip. After analyzing the literature, it can be concluded that for a metallized layer, a thickness of 500-900 nm is sufficient. This element is very important in the operation of the device, as it affects the accuracy of the transfer of the result to the electronic device [4].

Table 2 presents a comparison of the two most common metals to create this layer.

Table 2. Analysis of metallization materials for MFC

Characteristic	**Copper**	**Gold + Titan**
Electrical conductivity, S/m	$5{,}96 \cdot 10^7$	$4{,}1 \cdot 10^7$
Resistance, μΩ*cm	1,7	2,3
Specific thermal conductivity, W/(m*K)	398	318
Economic component, rub/g	0,375	2717

By all measures, most of all, for a given chip element, copper is suitable. This metal has the highest conductivity and less resistance than gold. Copper is the most profitable from the economic point of analysis, it is much cheaper. Also, it does not need, in contrast to gold, an additional component to improve adhesion. The most important disadvantage of copper is high oxidability, but this disadvantage will not have a strong impact on the work, since the conductive element is easily replaced.

4. The choice of the method of forming a nanoscale conductive layer length microfluidic chip

One of the key operations is to create a conductive layer for transferring analysis data to an electronic device. There are several ways to create metallic nanoscale copper layers on the surface of the base: magnetron sputtering, laser ablation, thermal evaporation, arc deposition, schemes of which are shown in Figure 2.

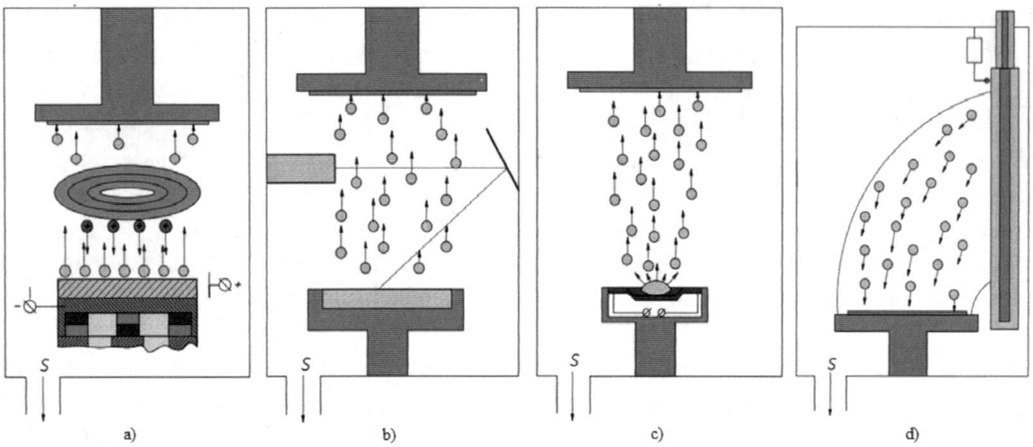

Figure 2. Application scheme: a) Magnetron deposition. b) Laser evaporation. c) Thermal evaporation. d) Arc deposition.

The result of the analysis of these methods is presented in Table 3:

Mode	Sedimentation rate, μm /h	Operating pressure, Pa	Material utilization	Substrate temperature, K	Adhesion strength, MPa
Magnetron sputtering	10^{-2}-$5x10^{-1}$	10^{-5}-10^{-2}	Low	800	Until 150
Laser evaporation	0,1-1	10^{-7}-10^{-5}	Low	400	Until 100
Thermal evaporation	0,1-1	10^{-5}-10^{-2}	High	1000	Until 100
Arc deposition	0,1-50	10^{-5}-10^{-2}	Medium	500	Until 310

Table 3. Comparison of film application methods

Based on the available data on the methods required by the parameters of the conductive layer, as well as on the available equipment, the most appropriate method is the magnetron sputtering method, which is implemented on the UVN-1M installation.

5. Technology development and equipment selection

Based on the analysis of suitable materials, the microfluidic chip manufacturing technology was developed. It can be divided into three stages.

The first stage involves the creation of a conductive base. To do this, clean the polymer plate. This will reduce the substrate roughness, due to contamination on the surface, and also to avoid deterioration of adhesion, and contamination of the film structure.

This operation will be performed on a low-pressure plasma installation using ion-plasma etching. The estimated process time is 30 minutes.

The formation of a conductive layer is a key operation for the first stage of manufacturing a microfluidic chip. This operation is performed on a vacuum deposition unit UVN-1M by magnetron sputtering at an operating pressure of $5x10^{-1}$ Pa in the working gas medium (argon). The approximate pumping time is 10 minutes.

In the second stage, an analysis block is created. As in the first stage, it begins with the cleaning of the polymer plate. Then, to create a system of capillaries, the method of photolithography was chosen. A positive photoresist is applied to the cleaned plate using a centrifuge to further etch certain areas of the polymer.

Next is the formation of the mask. This operation consists of several stages, and is necessary for giving the photoresist a given structure for further etching. First of all, it is necessary to dry the photoresist using a thermostat, for the best adhesion to the surface of the plate. The operation time is 30 minutes. Then the photomask is combined with the sample.

After combining the photomask and the surface of the polymer plate with a photoresist applied to it, the photoresist is illuminated through the photomask using ultraviolet radiation. After the photoresist is exposed, the photomask is removed. Then a photoresist is developed, during which the lighted part of the photoresist is removed, leaving a non-lighted photoresist on the substrate surface of the required topology. Operation time 30 s.

After creating the mask of the required pattern, areas of the polymer plate are etched, which are not hidden under the photoresist layer.

The photoresist plate is placed in the chamber of the etching unit, where the polymer is etched by reactive ion etching in CHF_3 medium, for 60 minutes.

After forming the desired structure, remove the layer of the remaining photoresist. Alkali solution is best for its removal. Alkali solution concentration: 5%.

After removing the photoresist, we obtain the necessary structure in the polymer plate - a polymer stamp.

The third stage is packaging. After creating the first and second main parts of the microfluidic chip housing, you can proceed to their connection. The simplest method of bonding will be diffusion glue, since this substance must be transparent and also have the properties of a dielectric.

After applying the glue around the perimeter of the polymer plate, it should be left in the air for 40-60 seconds, then attach the polymer template. Then, in the same way, a sealing cap is attached. This operation, for greater accuracy of alignment, it is recommended to make along the guides. After bonding both parts, leave the chip for a day, for complete solidification.

An illustration of this technology is presented in Figure 3.

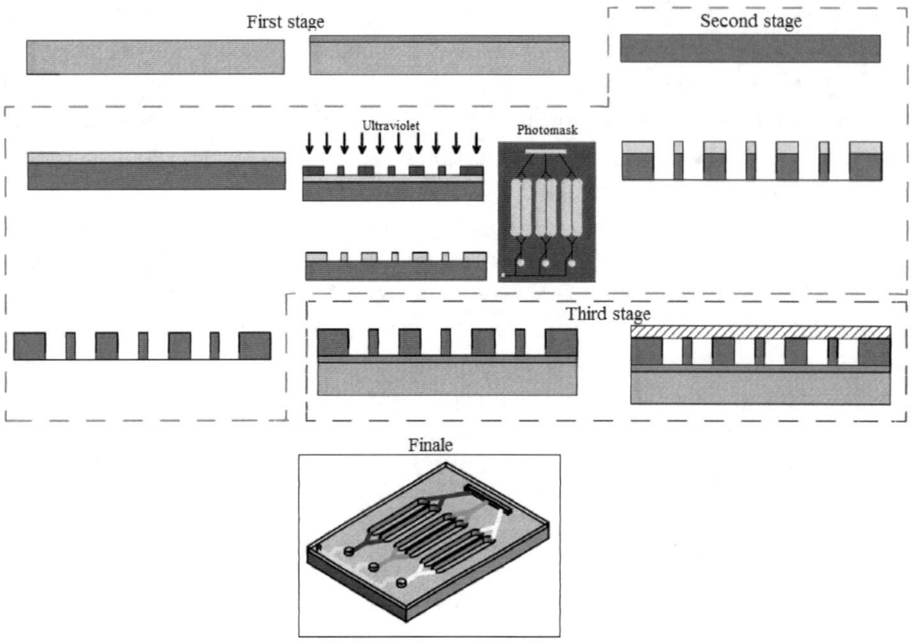

Figure 3. Diagram of microfluidic chip manufacturing technology

The developed technology is a sequence of operations, enlarged in stages - there are three of them. As a result of the implementation of this technology, it is possible to create a microfluidic chip, which allows determining the main components of human sweat: ammonia, iodine and urea.

6. Setting up and conducting an experiment to obtain the dependence of resistance on process parameters.

To study the dependence of the dimensional and electrical characteristics of the conductive layer, a series of experiments was carried out.

After studying the materials of the base, the conductive layer, as well as the methods of applying a thin film coating, the following components were selected:

- silicon substrates, since further work with the microfluidic chip will be carried out on the basis of silicon, the experiment should be carried out immediately on this material;
- copper, to create a conductive layer;
- installation supporting the magnetron sputtering method.

Pre-cleaned substrates, using an alcohol solution, were placed separately in a vacuum chamber. The spraying was carried out according to the given data of time and distance from the substrate to the source. Photos of the samples, as well as the values of time and distance are shown in Figure 4.

IOP Conf. Series: Materials Science and Engineering **781** (2020) 012022 doi:10.1088/1757-899X/781/1/012022

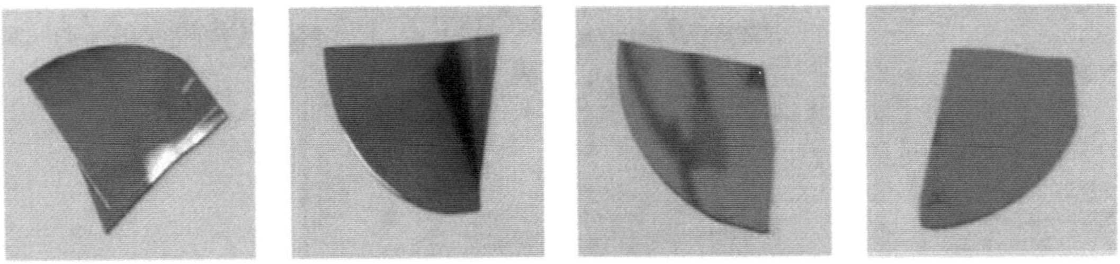

1) Time: 3 min Distance: 30 mm | 2) Time: 6 min Distance: 30 mm | 3) Time: 3 min Distance: 50 mm | 4) Time: 6 min Distance: 50 mm

Figure 4. Photographs of the obtained samples

To obtain the parameters of the electrical network conductivity and resistance, an immitance meter was used, also called an RLC meter. The measurement was carried out according to the scheme shown in Figure 5.

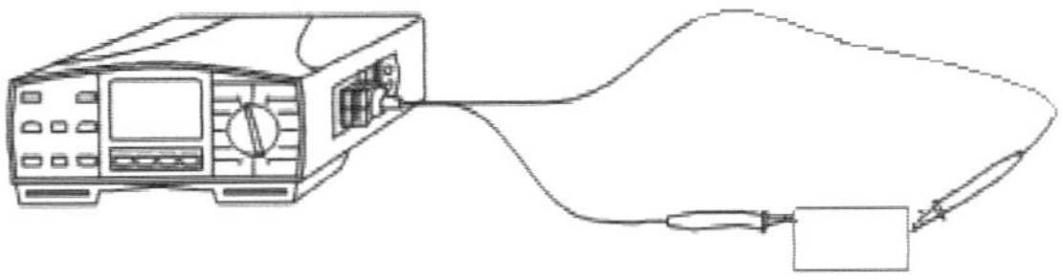

Figure 5. Resistance measurement circuit

After all measurements, the following melon
for sample 1 (3 min, 30 mm) the resistance is 4583.66 mOhm;
for sample 2 (6 min, 30 mm) resistance is 28x103 mOhm;
for sample 3 (3 min, 60 mm) the resistance is 991.6 mOhm;
for sample 4 (6 min, 60 mm) the resistance is 7.61 x 103 mOhm.

According to the measurement data on the RLC meter, a mathematical model was built of the dependence of the resistance of the copper coating on the application time and the distance from the substrate to the source:

$$R = 10304 + 7516 \cdot t - 6003 \cdot h - 4207 \cdot t \cdot h \quad (1)$$

where h is the distance from the source to the substrate, t is the process time.

After analyzing the obtained mathematical model of the process, we can conclude that both the distance from the source to the substrate and the deposition time affect quite significantly. And also, the interaction of these two factors has a significant impact on the value of resistance. The mathematical model confirmed its adequacy, the experiments were reproducible.

7. Conclusions

In the course of this work, an analysis of the base materials and the current-conducting layer of the microfluidic chip was carried out. The most suitable materials are: for the case - polymer or glass, for the conductive layer - copper. Conductive layer can be applied in vacuum using magnetron deposition.

The proposed technology, at the moment is a theoretical development, and is being tested for feasibility in MSTU. N. E. Bauman.

References

[1] Beebe D J 2015 Microfluidics: theory and applications 120-56

[2] Bissacco G 2016 Precision manufacturing methods of inserts for injection molding of microfluidic systems 10-16
[3] Evstrapov A A 2011 Microfluidic chips for biological and medical research 109-47
[4] Nisan A 2016 Microfluidic modules: applications and production technologies 11-19

AUTHOR INDEX